Lecture Notes in Economics and Mathematical Systems

376

W0051435

K.-H. Jöckel   G. Rothe   W. Sendler (Eds.)

Bootstrapping and Related Techniques

Proceedings of an International Conference
Held in Trier, FRG, June 4-8, 1990.

Springer-Verlag
Berlin Heidelberg New York
London Paris Tokyo
Hong Kong Barcelona
Budapest

Editors

PD Dr. Karl-Heinz Jöckel
Bremen Institute for Prevention Research
and Social Medicine (BIPS)
Grünenstraße 120, W-2800 Bremen, FRG

Prof. Dr. Günter Rothe
Hauptverband der gewerblichen Berufsgenossenschaften e. V.
Alte Heerstraße 111, W-5205 Sankt Augustin, FRG

Prof. Dr. Wolfgang Sendler
University of Trier, Abteilung für angewandte Mathematik
Postfach 3825, W-5500 Trier, FRG

ISBN-13: 978-3-540-55003-7   e-ISBN-13: 978-3-642-48850-4
DOI: 10.1007/978-3-642-48850-4

Typesetting: Camera ready by author

42/3140-543210 - Printed on acid-free paper

Editorial

Although methods like the Bootstrap, the Jackknife, Monte-Carlo-Tests and resampling techniques in general have been known for many years (a famous example is Gosset's ('Student') simulation of the bivariate sample correlation coefficient), a spread of these techniques may be realized with the increasing availibility of modern computer technology in all areas of science. Especially Efron's 1979 paper on the bootstrap had a high impact on this development. It should, however, be mentioned that in some fields like spatial analysis application of Monte-Carlo-Tests had been in use since the sixties. In parallel to the availibility of modern computer technology there has been an increasing interest in the theoretical aspects of these techniques making them nowadays an established element of computational statistics. Subsequently more and more applied researchers tend to apply these techniques although the theoretical applicability sometimes seems to be in question.

This was the starting situation when during the 1987-meeting of the German Region of the International Biometric Society held in Trier the editors of this volume decided to organize an international conference on the subject. The idea of organizing such a meeting was approved by the working group 'Computational Statistics' of the Biometric Society, German Region and prepared in cooperation with the working group 'Statistical Analysis Systems' of the German Association for Medical Documentation and Statistics (GMDS). This conference was designed to bring together methodological experts and practitioners and to initiate interaction between these groups. Finally the conference was held in Trier (in order to have at least two observations) June 4 - 8 1990 attracting 110 scientists. 14 invited and 35 contributed papers covering a broad spectrum of topics were included in the final program. 30 selected papers are presented in this volume. Each paper has been reviewed by at least two referees.

The book is organized in 9 sections roughly reflecting the section titles of the conference, where certain classification problems could not completely be avoided. The editors feel that the variety of papers and subjects correctly reflects the present state of research in this area from mathematically sophisticated limit theorems to the more technical end of Monte-Carlo as well as specific applications. Thus the editorial and reviewing policy did not only focus on methodological consistence but also on relevance in view of applied aspects. Since the editors felt that besides all purely scientific aspects the human sight should not be forgotten, we were glad that Mark Johnson supplied us with the historiography of his work presented as an invited paper at the conference (which by now has appeared in Technometrics). At this point we would like to acknowledge the help of all reviewers that are listed at the end of the book. Further thanks are due to the Deutsche Forschungsgemeinschaft, the Government of the State of Rhineland Palatinate and the following enterprises for financial support of the conference, resp. this proceedings volume: CIBA Geigy, Gödecke AG, Landesbank Rheinland-Pfalz, Schwarz-Pharma AG, Bayer AG, Leverkusen, Schering AG, Berlin, Merz u. Co. GmbH, Frankfurt, Canfor GmbH, Köln.

We would also like to thank Dr. H.-D. Keller from Trier for his consistent help in organizing the conference and handling all financial aspects.

Last but not least we would like to thank our secretaries D. Selter and M. Titze for doing a big job in assisting us to organize the conference. Without their aid the preparation of this volume would not have been possible.

Bremen and Trier, June 1991   K.-H. Jöckel, G. Rothe, W. Sendler

Contents

Random Number Generation (1)

PRINCIPLES FOR GENERATING
NON–UNIFORM RANDOM NUMBERS

Ulrich DIETER

1. Introduction: The Generation of Uniform Random Variables

In many simulations random numbers from a given distribution function $F(x)$ with density $f(x) = F'(x)$ are necessary. If $f(x) = 1$ for $0 \leq x \leq 1$ *uniformly distributed random numbers* are needed. For this purpose the *linear congruential method* is widely used: A sequence of integers is initialized with a value z_0 and continued as

$$z_{i+1} \equiv az_i + r \pmod{m}, \qquad 0 \leq z_i < m \quad \text{for all} \quad i.$$

The fractions $u_i = z_i/m$ are the derived pseudo-random numbers in the interval $[0,1)$. The constants m, the modulus, a, the multiplicator, r, the increment, and z_0, the starting number are suitably chosen non–negative integers. Three choices of m, a and r are common on most computers:

1. $r = 0$, $m = 2^E$, $a \equiv 5 \pmod 8$ and $z_0 \equiv 1 \pmod 4$. All $z_i \equiv 1 \pmod 4$ are generated.
2. $r = 0$, $m = p$, p prime, a a primitive root mod p. All $z_i = 1, \ldots, p-1$ are generated.
3. $\gcd(r,m) = 1$, $m = 2^E$, $a \equiv 1 \pmod 4$. All integers $0, 1, \ldots, 2^E - 1$ are generated.

For selecting *good* random number generators one has to study the distribution of the k-tuplets $P_k = (u_{i+1}, u_{i+2}, \ldots, u_{i+k})$. Geometrically theese P_k may be considered as points of a lattice G in the k-dimensional hypercube $[0,1)^k$. The lattice points can also be seen as intersection points of k sets of parallel hyperplanes. Consequently, the following questions may be raised:

(i) Determine the minimal number N_k^* of parallel hyperplanes on which all points P_k lie.
(ii) Determine the maximal distance D_k^* of parallel hyperplanes on which all points P_k lie.

Question (i) was asked by G. Marsaglia in his well known article "Random numbers fall mainly in the planes". There he derived upper bounds for N_k^* using Minkowski's *convex body theorem*. The 'wave numbers' $W_k^* = 1/D_k^*$ were introduced by Coveyou and MacPherson in their "Fourier Analysis of random number generators". Their algorithm for calculating W_k^* was simplified by D.E. Knuth in the first edition of Volume II of his book *The Art of Computer Programming* in his chapter on the *Spectral Test*.

The calculation of both quantities is based on a general procedure to determine non-zero vectors of shortest *length* in a lattice. For the determination of N_k^* the ℓ_1–norm is used, and for D_k^* the Euclidean norm is appropriate. The author's algorithm of 1973 gave exact values for both quantities; no exact values for N_k^* were known before. D.E. Knuth included a variant of our algorithm in the second edition of Volume II of his *Art of Computer Programming*. A completely different approach was proposed by Lovacs and the brothers Lenstra, called the L^3-algorithm. In the case of the Euclidean norm the final search can be shortened by an idea of Finke and Pohst.

For any sequence $\{u_i\}$ of $[0,1)$-uniformly distributed random numbers the local deviation

$$\Delta_k(\mathbf{s},\mathbf{t}) = \frac{\#\{U_i = (u_{i+1}, \ldots, u_{i+k}) \mid s_j < u_{i+j} \leq t_j, \ j = 1, \ldots, k\}}{\#\{U_i = (u_{i+1}, \ldots, u_{i+k}), \ j = 1, \ldots, k\}} - \prod_{j=1}^{k}(t_j - s_j)$$

and their largest value, the (global) discrepancy

$$\Delta_k = \sup \{|\Delta_k(\mathbf{s},\mathbf{t})| \ \mid \ 0 \leq s_j \leq t_j < 1, j = 1, \ldots, k\}$$

are of great importance. For example, for the calculation of k–dimensional integrals by Monte–Carlo methods the difference of the integral and its approximation by a Riemann sum is bounded by the discrepancy Δ_k multiplied by the variation of the function $V(f)$ (in the sense of Hardy and Krause). Since the variation of the function is fixed, the discrepancy has to be as small as possible.

No methods are known for calculating the discrepancy in dimension greater than two. The author derived a lower bound in 1973, and H. Niederreiter found an upper bound in 1978. Even these bounds are difficult to calculate.

In dimension 2, i.e. in the case of pairs, all three quantities can be calculated by the Euclidean algorithm for the period length n and a. n is equal to $m/4$ if $m = 2^E$ and $r = 0$ and $n = m$ in the two other cases. Define a sequence $\{m_i\}$ by

$$m_0 = n, \quad m_1 = a, \quad m_{i-1} = a_{i-1}m_i + m_{i+1}, \quad i = 1.2, \ldots, \quad \text{where } a_{i-1} = \left\lfloor \frac{m_{i-1}}{m_i} \right\rfloor,$$

and $\lfloor x \rfloor$ is the *integer* function. Associated is the sequence

$$p_0 = 0, \quad p_1 = 1, \quad p_{i+1} = a_{i-1}p_i + p_{i-1}, \quad i = 1.2, \ldots .$$

Then

$$N_2^* = \min_i \{m_i + p_i\} \quad \text{and} \quad W_2^* = 1/D_2^* = \min_i \sqrt{m_i^2 + p_i^2} . \qquad (1,1)$$

Finally, for the discrepancy the following rather sharp bounds hold

$$\frac{1}{4n} \max\{a_i \mid 0 \le i \le t\} \le \Delta_2 \le \frac{1}{4n} \left(\sum_{i=0}^{t} a_i + 2 \right) . \qquad (1,2)$$

In a paper of L. Afflerbach and R. Weilbächer and in a forthcoming article of U. Dieter methods for calculating exact values of the discrepancy are presented. The numerical values differ only slightly from the upper bound in (1,2). The three expressions in (1,1) and (1,2) are easy to calculate.

If the dimensions become larger, the number of hyperplanes, on which the lattice points P_k lie, decrease considerably. Therefore a different procedure was proposed by D.E. Knuth (1969): a sequence of integers z_i is initialized to $(z_0, \ldots, z_{r-1}) \ne (0, \ldots, 0)$ and up-dated by

$$z_i \equiv a_1 z_{i-1} + a_2 z_{i-2} + \cdots + a_r z_{i-r} \pmod{p} \quad 0 \le z_i < p. \qquad (1,3)$$

Here the factors a_i are given integers, and for the modulus p only prime numbers are considered. Again, the fraction $u_i = z_i/p$ are taken as random samples from the interval $[0, 1)$.

Since there are only p^r possible r-tuplets $(z_k, z_{k+1}, \ldots, z_{k+r-1})$ and $(0, 0, \ldots, 0)$ must not occur, the period length of (1,3) is at most $p^r - 1$. It can be shown that this maximum period length of $p^r - 1$ may in fact be achieved for suitable choices of the factors $a_1, a_2, \ldots, a_r$. This means that *all* r-tuplets $(z_k, \ldots, z_{k+r-1}) \ne (0, 0, \ldots, 0)$ must occur resulting in perfectly uniform distributions of the $u_i = z_i/p$, the pairs (u_i, u_{i+1}), the triplets (u_i, u_{i+1}, u_{i+2}) (if $r \ge 3$) etc. In other words, for all dimensions $k \le r$ the hypercube $[0, 1)^k$ is now evenly filled. For dimensions $k > r$ the quantities N_k^* and D_k^* can be calculated exactly, since the points P_k are again points of a lattice. The numerical values show that a sequence of type (1,3) behaves as good in dimension $k \times r$ as a linear congruential generator behaves in dimension k. Furthermore, reduced bases (in the sense of Minkowski) can be determined which show how *good* the specific generator behaves. For this see the dissertation of my former student Grube of 1973 and the recent publications of Grothe.

Recently, other recipes for generating pseudo-random numbers have been proposed by J. Lehn and his co-workers. They are discussed in his contribution.

2. Direct Generation of Non–Uniform Random Variables

If the distribution function is different from the uniform one, uniformly distributed random numbers have to be transformed. Some well–known methods are applied:

Inversion Methods: $X = F^{-1}(U)$ has the distribution function $F(x)$ if F^{-1} is the inverse function of F and if U is uniformly distributed in $[0,1)$. Unfortunately, only a small number of distributions are known for which *simple* inversion functions exist: the *exponential* distribution $F(x) = 1 - e^{-\lambda x}$ and the *Cauchy* distribution $f(x) = F'(x) = c/\pi(1+c^2(x-b)^2)^{-1}$ are the most prominent examples:

$$X \leftarrow -\frac{1}{\lambda}\ln U \qquad \text{samples from the } exponential \ distribution;$$

$$X \leftarrow b + \frac{1}{c}\tan(\pi U) \qquad \text{samples from the } Cauchy \ distribution.$$

Furthermore, $F(x) = x^n$ can be generated as

$$X \leftarrow U^{1/n} \qquad \text{or} \qquad X \leftarrow \max\{U_1, \ldots, U_n\}$$

Hence, picewise polynomial approximations of $F^{-1}(U)$ can always be constructed. However, since they need large tables of constants the evolving algorithms are often slow.

In the special case of $F(x) = x^2$ an interesting procedure was invented by J. Ahrens, called *Bit Scrambling*:

1. Generate U; let $U = .u_1 u_2 \ldots$ be the binary representation of U.
   Let u_j be the first bit that is one: $u_1 = \cdots = u_{j-1} = 0, u_j = 1$.
2. If $j = 1$ set $X \leftarrow U$. Otherwise set $X \leftarrow .u_{j+1}u_{j+2}\ldots u_{2j-1}1u_{2j}u_{2j+1}\ldots$

It is now proved that this algorithm simulates the maximum of two independent $(0,1)$-deviates. If $j = 1$ (probability $1/2$) then $1/2 \leq X = U < 1$. This contributes the partial probability density function $f_1(x) = (1/2) \times 2 = 1$ in $[1/2, 1)$. If $j = 2$ (probability $1/4$) then U is either $U = .010\ldots$ giving $X = .01\ldots$ or $U = .011\ldots$ resulting in $X = .11\ldots$. Hence the second partial density is $f_2(x) = (1/4) \times 2 = 1/2$ in the intervals $[1/4, 1/2)$ and $[3/4, 1)$. In general, $f_j(x) = 2/2^j$ in the 2^{j-1} intervals in which the j-th bit of the abscissa x is a one (and $f_j(x) = 0$ elsewhere). It follows that

$$f(x) = \sum_{j=1}^{\infty} f_j(x) = \sum_{j=1}^{\infty}(2/2^j \mid j\text{-th bit of } x \text{ is } 1) = 2x$$

and consequently $F(x) = x^2$.

In their important article "The complexity of random number generation" D.E. Knuth and A.C. Yao [1976] discuss the generation of deviates from arbitrary distributions given a source of random *bits*. They produced a slightly different algorithm for sampling from $F(x) = x^2$.

The *Box–Muller* method

$$X' \leftarrow \sqrt{-2\ln U'}\sin(2\pi U''), \qquad X'' \leftarrow \sqrt{-2\ln U'}\cos(2\pi U'')$$

transforms two independent $(0,1)$–uniformly distributed random variables U' and U'' into two independent *standard normal variables* X' and X''.

For the *Ratio of Uniforms* one needs the transformation

$$X' = m + \frac{sU'' - t}{U'^{1/k}}, \qquad X'' = U'^{1+1/k}.$$

Since the Jacobian

$$\frac{\partial(X', X'')}{\partial(U', U'')} = \begin{vmatrix} -\dfrac{sU'' - t}{kU'^{1+1/k}} & \dfrac{s}{U'^{1/k}} \\ (1 + 1/k)U'^{1/k} & 0 \end{vmatrix} = s(1 + 1/k)$$

is constant, the variables X', X'' are again uniformly distributed in their domain D. It is bounded by the following curves:

The line $U' = 0, U'' \in [0,1)$ corresponds to $-\infty < X' < \infty$, $X'' = 0$.

The line $U' = 1, U'' \in [0,1)$ corresponds to $X' \in [m-t, m-t+s]$, $X'' = 1$.

The line $U' \in [0,1), U'' = 0$ corresponds to $X' \in (-\infty, m-t]$, $X'' = (t/(m-X'))^{k+1}$.

The line $U' \in [0,1), U'' = 1$ corresponds to $X' \in [m-t+s, \infty)$, $X'' = ((s-t)/(X'-m))^{k+1}$.

This means that the transformed area is bounded by the *table montain function*

$$h(x) = \begin{cases} \left(\dfrac{t}{m-x}\right)^{k+1} & x \in (-\infty, m-t] \\ 1 & x \in [m-t, m-t+s] \\ \left(\dfrac{s-t}{x-m}\right)^{k+1} & x \in [m-t+s, \infty) \end{cases}$$

and the X–axes. It will be used as a *hat* function in the next section.

In 1951 John von Neumann published an ingenious method for sampling from the exponential distribution which was based solely on *comparisons of uniform deviates*. After the author had modified this algorithm in order to obtain samples from the center $[-\sqrt{2}, \sqrt{2}]$ of the standard normal distribution, G.E. Forsythe remembered that v. Neumann had once hinted at possible generalizations. There was also a remark in that direction in his notes which he had taken during one of v. Neumann's lectures. This motivated Forsythe to take up the subject again, and shortly he came up with a method which is applicable to all probability density functions of the form $\alpha \exp(-G(x))$.

1. Generate U. Set $X \leftarrow a + (b-a)U$ and $T \leftarrow G(X)$.
2. Generate $U_1, \dots, U_K$. K is determined by the condition
   $T \geq U_1 \geq \cdots \geq U_{K-1} < U_K$ (If $T < U_1$, then $K = 1$).
3. If K is even, reject X and go back to (1).
4. If K is odd, return X as a sample from (6,2).

For $G(x) = x$ one obtains samples from the exponential distribution in the interval $(0,1)$, and for $G(x) = x^2$ from the standard normal distribution in $(0, \sqrt{2})$.

So far the method can only be applied to intervals $a \leq x \leq b$ for which $0 \leq G(x) \leq 1$. For variables from other intervals $(b, b+d)$ the algorithm is now extended by changing the probability density function $\alpha \exp(-G(x))$ into $\alpha \exp(-G(b)) \exp(-(G(x) - G(b))) = \alpha^* \exp(-G^*(x))$. The new function $G^*(x) = G(x) - G(b)$ has the property $G^*(b) = 0$, and it works as long as $0 \leq G^*(x) \leq 1$. Continuing in this way a procedure emerges for sampling from the entire interval $[0, \infty)$.

In this manner extremely fast methods for sampling from the exponential and from the normal distribution can be constructed. However, these methods need tables of constants. For details see the papers of G. Forsythe (1972) and J.H. Ahrens and U. Dieter (1973). For other distributions applications of the comparison method have been tried. However, if the parameters of these distributions are changed, new tables have to be constructed. Consequently, the emerging algorithms are not very efficient.

3. Acceptance–Rejection Methods for Sampling of Random Variables

During the last twenty years it became clear that this method, also introduced by J.v.Neumann, is the most adaptable method for sampling from complicated distributions. It works as follows:

Let $f(x)$ be a given probability density, and let $h(x)$ be a function such that $f(x) \leq h(x)$ within the range of $f(x)$. If the integral of $h(x)$ over this range is a finite number α, then $g(x) = h(x)/\alpha$ is a probability density function, and the following procedure is valid:

1. Take a random sample X from the distribution with probability density $g(x) = h(x)/\alpha$.
2. Generate a uniform random deviate U between zero and one. If $U \leq f(X)/h(X)$, accept X as a sample from the distribution $f(x)$. Otherwise reject X and go back to Step 1.

The ease of the method depends on the following properties of the hat function $h(x)$:

A. One has to select a hat function $h(x)$ from which it is easy to sample.
   Examples are normal, double–exponential and triangular densities.
B. The parameters of the hat function have to be determined in such a way that the area α below $h(x)$ becomes minimal.

It will be shown that optimal hat functions can be calculated by analytical methods. Some of the published algorithms use hat functions which are far away from the optimal ones. It is assumed that the hat function $h(x)$ touches $f(x)$ at *two* points L (left) and R (right) where $L < R$. Furthermore, we suppose that $h(x)$ depends on two parameters, called m and s. Thus we demand that

$$f(L) = \alpha\, g(L; m, s)\,, \quad f(R) = \alpha\, g(R; m, s) \tag{3,1}$$

and $f(x) \leq \alpha\, g(x; m, s)$ for all other x. Since L and R are local maxima of $f(x)/g(x; m, s)$, we have the necessary conditions

$$\frac{f'(L)}{f(L)} = \frac{g'(L; m, s)}{g(L; m, s)} \quad \text{and} \quad \frac{f'(R)}{f(R)} = \frac{g'(R; m, s)}{g(R; m, s)}. \tag{3,2}$$

If L and R are uniquely determined, they should satisfy the sufficient conditions

$$\frac{f''(L)}{f(L)} < \frac{g''(L; m, s)}{g(L; m, s)} \quad \text{and} \quad \frac{f''(R)}{f(R)} < \frac{g''(R; m, s)}{g(R; m, s)}. \tag{3,3}$$

Otherwise, the first derivative of $\ln\big(f(x)/g(x; m, s)\big)$ has to be discussed in detail.

Equations (3,1) and (3,2) are four equations for the determination of L, R, m, s and α. Assuming that L, R, and m can be expressed as functions of s, we have to minimize

$$\alpha(s) = \frac{f\big(L(s)\big)}{g\big(L(s); m(s), s\big)} = \frac{f\big(R(s)\big)}{g\big(R(s); m(s), s\big)}. \tag{3,4}$$

This leads to the necessary conditions

$$-\frac{d}{ds}\ln\alpha(s) = -\frac{d}{dL}\ln\frac{f(L)}{g(L; m, s)}\frac{dL}{ds} + \frac{d}{dm}\ln g(L; m, s)\frac{dm}{ds} + \frac{d}{ds}\ln g(L; m, s) = 0$$

$$-\frac{d}{ds}\ln\alpha(s) = -\frac{d}{dR}\ln\frac{f(R)}{g(R; m, s)}\frac{dR}{ds} + \frac{d}{dm}\ln g(R; m, s)\frac{dm}{ds} + \frac{d}{ds}\ln g(R; m, s) = 0\,.$$

In both equations the first expression after the equal–sign is zero by (3,2). Solving both equations for dm/ds, and comparing, yields the fundamental relation

$$\frac{d\ln g(L; m, s)}{dm}\frac{d\ln g(R; m, s)}{ds} = \frac{d\ln g(R; m, s)}{dm}\frac{d\ln g(L; m, s)}{ds} \tag{3,5'}$$

or, by observing $\frac{d}{dm}\ln g(L; m, s) = \frac{1}{g(L; m, s)}\frac{d}{dm}g(L; m, s)$

$$\frac{d\, g(L; m, s)}{dm}\frac{d\, g(R; m, s)}{ds} = \frac{d\, g(R; m, s)}{dm}\frac{d\, g(L; m, s)}{ds} \tag{3,5''}$$

(3,2), (3,4) and (3,5') or (3,5") contain five conditions for finding candidates L, R, m, s and α. Whether a solution will in fact lead to a local minimum of α has to be checked carefully in each special case.

We shall consider four examples of possible hat functions $h(x)$ that touch given probability densities $f(x)$ at two locations L and R.

Triangular Hat Functions. The first example deals with densities that can be enclosed in an isosceles triangle $h(x)$, and whose corresponding density $g(x) = h(x)/\alpha$ depends on the parameters m and s as follows.

$$g(x; m, s) = \frac{1}{s} - \frac{1}{s^2}|x - m| \qquad x \in [m - s, m + s].$$

Samples may be obtained as $X \leftarrow m + s(U_1 + U_2 - 1)$ where U_1 and U_2 are $[0,1)$–uniformly distributed. In this case the fundamental identity leads to

$$s = R - L.$$

With its help all constants L, R, s, m and α can be determined.

Double Exponential Hat Functions. Our second example of a hat function with two points of osculation, L and R, has infinite range; it is the double exponential (or Laplace) distribution with density

$$g(x; m, s) = \frac{1}{2s} \exp\left(-\frac{|m - x|}{s}\right).$$

Samples are obtained as $X \leftarrow m + TsE$ where E is a standard exponential deviate and T a random sign $\pm$. The fundamental identity (3,5") yields

$$2s = R - L.$$

Student-t Hat Functions. Its probability density function is given as

$$t_n(x) = \frac{c_n}{s}\left(1 + \frac{(x - m)^2}{s^2 n}\right)^{-\frac{n+1}{2}} \qquad \text{for } -\infty < x < \infty, \, n \geq 1,$$

where

$$c_n = \frac{1}{\sqrt{n}\, B(n/2, 1/2)} = \frac{\Gamma\big((n+1)/2\big)}{\sqrt{n\pi}\, \Gamma(n/2)}.$$

The t-family contains the Cauchy distribution for $n = 1$ and the normal distribution for $n \to \infty$ as extreme cases. For $n = 2$ and $n = 3$ special sampling methods are available: If U denotes a $(0,1)$–uniform random variable, then $X \leftarrow (U - 1/2)/(\sqrt{(U - U^2)/2})$ samples from t_2 and values from the t_3-distribution can be generated efficiently by the ratio-of-uniforms method of Kinderman/Monahan (1977) and (1980) which will be discussed later on.

Now the fundamental identity (3,5') yields

$$s^2 = (R - m)(m - L)$$

As before, L, R, s, m and α can be determined explicitly.

For the **Ratio of Uniforms Method** the *table mountain function* is taken as a hat function. Assume that $f(x)$ is a density and $f = max_x f(x)$. To simplify the notation we use $\bar{f}(x) = \frac{1}{f}f(x)$. For applying the acceptance–rejection method $\bar{f}(x) \leq h(x)$ must hold. This means $(m -$

$x)^{k+1}\bar{f}(x) \leq t^{k+1}$ for $x \leq m - t$ and $(x - m)^{k+1}\bar{f}(x) \leq (s - t)^{k+1}$ for $x \geq m - t + s$. Define
$v_- = \inf\{(x - m)(\bar{f}(x))^{1/(1+k)} \mid x \leq m - t\}$ and $v_+ = \sup\{(x - m)(\bar{f}(x))^{1/(1+k)} \mid x \geq m - t + s\}$
Our assumptions mean that v_- and v_+ are finite and $t = -v_-$, $s = v_+ - v_-$ are possible choices
for s and t. Sampling may be carried out in the following way.

1. Generate $[0,1)$–uniform random numbers U and V and set
   $X \leftarrow m + (sV - t)U^{-1/k}$, $Y \leftarrow U^{1+1/k}$.
2. If $Y = U^{1+1/k} \leq \bar{f}(X) = \bar{f}(m + (sV - t)U^{-1/k})$ return X as a sample from $f(x)$.
   Else go to 1.

The special case $m = 0$, $s = v_+ - v_-$, $t = -v_-$, $k = 1$ was introduced by Kinderman and
Monahan in 1977 in the following form.

1. Generate $[0,1)$–uniform random numbers U and V and set
   $U \leftarrow (\max_x f(x))^{1/2}U$, $V \leftarrow v_- + (v_+ - v_-)V$.
2. If $U^2 \leq f(V/U)$ return $X \leftarrow V/U$. Else go to 1.

So far optimal constants m, s, t have not been constructed. Again one has to minimize the
area $s(1 + 1/k)$ of D, i.e. the quantity s. Assume that $\bar{f}(x)$ touches $h(x)$ at a point $L < m - t$
and at $R > m - t + s$. This means $t = \max\{(m - x)(\bar{f}(x))^{1/(1+k)} \mid x \leq m - t\}$ and $s - t = \max\{(x - m)(\bar{f}(x))^{1/(1+k)} \mid x \geq m - t + s\}$ or

$$t = (m - L)(\bar{f}(L))^{1/(1+k)} \qquad s - t = (R - m)(\bar{f}(R))^{1/(1+k)}.$$

Addition leads to

$$s = (m - L)(\bar{f}(L))^{1/(1+k)} + (R - m)(\bar{f}(R))^{1/(1+k)}.$$

The derivative of s with respect to m becomes $\frac{ds}{dm} = (\bar{f}(L))^{1/(1+k)} - (\bar{f}(R))^{1/(1+k)}$ since the two
other expressions are zero by the optimality of L and R. Hence

$$\bar{f}(L) = \bar{f}(R), \quad i.e. \quad f(L) = f(R)$$

is a necessary condition for an optimum. This will determine L, R, m, t and s.

4. Sampling from Discrete Distributions

Inversion. Let a discrete distribution be given by its probabilities p_k and its cumulative distribution function be $P_k = \sum_{i=1}^{k} p_i$. Then generate a $[0,1)$–uniformly distributed random number
U and ask for which K $P_{K-1} < U \leq P_K$ holds. Then the integer K is accepted with the desired
probability p_K.

The method is popular in hand simulations. However, for the use on computers further considerations are necessary. One has to check whether the sum of the p_k is equal to 1. This is especially
important for infinite discrete distributions where the distribution has to be truncated as soon as
p_k lies below the accuracy of the machine. After such an adjustment the inversion method can
still be programmed in many different ways depending on the order in which the P_K are compared
with U until the unique value K satisfying $P_{K-1} < U \leq P_K$ is found. For this *searching procedures*
have to be used for which an extensive literatur exists.

There is one stricking example where this method works for a theoretical distribution. The
transformation

$$K \leftarrow \lfloor \log U / \log(1 - p) \rfloor$$

produces samples from the *geometric* distribution given by its probability function $p_k = p(1 - p)^k$
for $k = 0, 1, \ldots$.

For sampling from a *discrete uniform* distribution the transformation

$$K \leftarrow \lfloor nU \rfloor + 1$$

produces integers from 1 to n with equal probability $1/n$.

In a series of papers Walker ((1974a,b), (1977)) developed his *alias method* for sampling of random numbers from a discrete distribution. It relies on the following theorem.

Any discrete distribution with a finite number $n+1$ of mass points can be represented as an equiprobable mixture of n distributions, each of which has at most two mass points.

The theorem leads to the following sampling method:

1. Generate a discrete uniform deviate I between 1 and n.
   The two–point distribution I contains a *criminal i*, whose probability is q_i and its *alias a_i*.
2. Generate a uniform deviate U. If $U \leq q_i$, accept i, otherwise return its *alias a_i*.

For the proof of the theorem and for the construction of the *alias table* a simple lemma is needed.

Let b_i, $i = 1,\ldots,m+1$ be positive real numbers with sum m. Then

a. *There exist at least two b_i such that $b_i < 1$.*
b. *To $b_i < 1$ there exists a b_j such that $b_i + b_j \geq 1$.*

Proof. (a.) If only one $b_i < 1$, the sum of the remaining b_j is still greater m.

(b.) If $b_i+b_j < 1$ for all $j \neq i$, then $\sum_{j \neq i}(b_i + b_j) < m$, but $\sum_{j \neq i}(b_i + b_j) = (m-1)b_i+\sum_{k=1}^{m+1} b_k = (m-1)b_i + m > m$. This contradiction proves the lemma.

The last lemma is now used for the construction of the *alias table* in linear time. Since this method is important for the *bootstrap method*, an algorithm is presented.

0. Input p_i for $i = 1,\ldots,n+1$. Set $b_i \leftarrow np_i$, $i \leftarrow 0$, $j \leftarrow 0$.
1. Set $i \leftarrow i+1$. If $b_i > 1$ goto 1, else set $C \leftarrow i$.
2. Set $j \leftarrow j+1$. If $b_j \leq 0$ or $j = i$ goto 2.
3. If $b_j + b_i < 1$ and $j \leq n$ goto 2.
4. Set $a_i \leftarrow j$, $q_i \leftarrow b_i$, $b_j \leftarrow b_j + b_i - 1$, $b_i \leftarrow 0$.
5. If $j > C$ or $b_j > 1$ goto 7.
6. Set $i \leftarrow j$; goto 2.
7. Set $i \leftarrow C$, $j \leftarrow j - 1$. If $i < n$ goto 1.
8. Output i, its probability q_i and its alias a_i.

Recently, the ratio of uniforms was applied to discrete distributions. Ahrens and Dieter (1989) used it for the Poisson distribution and Stadlober (1989a,b) generated binomial and hypergeometric random variates by this procedure.

Further information might be obtained from the long *List of References* attached to this paper. On a few pages it is impossible to present a complete overview on the vast area of random number generation.

REFERENCES

Afflerbach, L. (1986), *The Sub-Lattice Structure of Linear Congruential Generators*, Manuscripta Math. **55**, 455–465.

Afflerbach, L. (1989), *Die Gütebewertung von Pseudo–Zufallszahlen–Generatoren aufgrund theoretischer Analysen und algorithmischer Berechnungen*, Mathematisch–Statistische Sektion **309**. Forschungsgesellschaft Joanneum, Graz, Austria.

Afflerbach, L. and Grothe, H. (1985), *Calculation of Minkowski-reduced lattice bases*, Computing **35**, 269-276.

Afflerbach, L. and Grothe, H. (1988), *The lattice structure of pseudo–random vectors generated by matrix generators*, J. Comp. Appl. Math. **23**, 127–131.

Afflerbach, L. and Weilbächer, R. (1989), *The exact determination of rectangle discrepancy*, Math. Comp. **53**, 343–354.

Ahrens, J.H. and Dieter, U. (1972), *Computer methods for sampling from the exponential and normal distributions*, Comm. ACM **15**, 873–883.

Ahrens, J.H. and Dieter, U. (1973), *Extension of Forsythe's method for random sampling from the normal distribution*, Math. Comp. **27**, 927–937.

Ahrens, J.H. and Dieter, U. (1974), *Computer methods for sampling from gamma, beta, Poisson and binomial distributions*, Computing **12**, 223–246.

Ahrens, J.H. and Dieter, U. (1980), *Sampling from binomial and Poisson distributions: A method with bounded computation times*, Computing **25**, 193–208.

Ahrens, J.H. and Dieter, U. (1982a), *Generating gamma variates by a modified rejection technique*, Comm. ACM **25**, 47–54.

Ahrens, J.H. and Dieter, U. (1982b), *Computer generation of Poisson deviates from modified normal distributions*, ACM Trans. Math. Software **8**,**2**, 163–179.

Ahrens, J.H. and U. Dieter (1988), *Efficient Table–free Sampling Methods for the Exponential, Cauchy and Normal Distributions*, Comm. ACM **31**, 1330–1337.

Ahrens, J.H. and U. Dieter (1989), *An Alias Method for Sampling from the Normal Distribution*, Computing **42**, 159–170.

Ahrens, J.H. and Dieter, U. (1990), *A convenient sampling method with bounded computation times for Poisson distributions*, American Journal of Mathematical and Management Sciences (to appear).

Atkinson, A.C. and Pearce, M.C. (1976), *The computer generation of beta, gamma and normal random variables*, J. Royal Statist. Soc. **A**, **139**, 431–461.

Barbu, G. (1982), *On computer generation of random variables by transformations of uniform variables*, Bulletin Mathématique de la Société des Sciences Mathématiques de la République Socialiste de Roumanie **26**, 129–139.

Beyer, W.A., Roof, R.B. and Williamson, D. (1971), *The Lattice Structure of Multiplicative Congruential Pseudo-Random Vectors*, Math. Comp. **25**, 345–360.

Coveyou, R.R. and MacPherson, R.D. (1967), *Fourier analysis of uniform random number generators*, J. ACM. **14**, 100–119.

Dagpunar, J. (1988), "Principles of Random Variate Generation," Clarendon Press, Oxford.

Deák, I. (1990), "Random Number Generators and Simulation," Akadémiai Kiadó, Budapest.

Devroye, L. (1986), "Non–uniform Random Variate Generation," Springer, New York.

Dieter, U. (1971), *Pseudo-random numbers: the exact distribution of pairs*, Math. Comp. **25**, 855–883.

Dieter, U. (1972), *Statistical Interdependence of pseudo-random numbers generated by the linear congruential method*, in "Applications of Number Theory to Numerical Analysis," ed. by S.K. Zaremba, Academic Press, Inc., New York and London.

Dieter, U. (1975), *How to calculate shortest vectors in a lattice*, Math. Comp. **29**, 827–833.

Dieter, U. (1979), *Schwierigkeiten bei der Erzeugung gleichverteilter Zufallszahlen*, Proceedings in Operations Research **8**, 249–272.

Dieter, U. (1982), *An alternate proof for the representation of discrete distributions by equiprobable mixtures*, J. Appl. Prob. **19**, 869 – 872.

Dieter, U. (1985), *Calculating Shortest Vectors in a Lattice*, Berichte der Math. Stat. Sektion im Forschungszentrum Graz **244**, 1-14.

Dieter, U. (1986), *Probleme bei der Erzeugung gleichverteilter Zufallszahlen*, in "L. Afflerbach und J. Lehn: Zufallszahlen und Simulationen," Teubner, Stuttgart, pp. 7–20.

Dieter, U. (1989), *Mathematical Aspects of Various Methods for Sampling from Classical Distributions*, Winter Simulation Conference 1989, Washington, D.C., 477–483.

Dieter, U. (1990), *Optimal acceptance–rejection methods for sampling from various distributions*, American Journal of Mathematical and Management Sciences (to appear).

Dieter U. (1991a), *Pseudo–Random Numbers: The Discrepancy in two Dimensions (to appear)*.

Dieter U. (1991b), *Criteria for Selecting good linear congruential generators (to appear)*.

Dieter U. and Ahrens, J.H. (1971), *An Exact Determination of Serial Correlations of Pseudo–Random Numbers*, Numer. Math. **17**, 101–123.

Dieter, U. and Ahrens, J.H. (1974), "Uniform Random Numbers," Inst. f. Math. Statistik, Technische Hochschule Graz.

Finke, U. and Pohst, M. (1985), *Improved methods for calculating vectors of short length in a lattice, including a complexity analysis*, Math.Comp. **44**, 463–471.

Fishman, G.S. (1978), "Priciples of Discrete Event Simulation," J. Wiley, New York.

Fishman, G.S. (1990), *Multiplicative congruential random number generators with modulus 2^β: An exhaustive analysis for $\beta = 32$ and a partial analysis for $\beta = 48$*, Math. Comp. 54, 331–344.

Forsythe, G.E. (1972), *Von Neumann's comparison method for random sampling from the normal and other distributions*, Math. Comp. 26, 817–826.

Grothe, H. (1988), *Matrixgeneratoren zur Erzeugung gleichverteilter Pseudozufallsvektoren*, Dissertation, TH Darmstadt.

Grube, A. (1973), *Mehrfach rekursiv erzeugte Zufallszahlen*, Dissertation Karlsruhe.

Kinderman, A.J. and Monahan, J.F. (1977), *Computer generation of random variables using the ratio of uniform deviates*, ACM Trans. Math. Software 3, 257–260.

Kinderman, A.J. and Monahan, J.F. (1980), *New methods for generating Student's t and gamma variables*, Computing 25, 369–377.

Knuth, D.E. (1981), "The art of computer programming, Vol. II: seminumerical algorithms," Addison–Wesley, Reading, Mass..

Kronmal, R.A. and Peterson, A.V. (1981), *A variant of the acceptance–rejection method for computer generation of random variables*, J. Amer. Statis. Assoc. 76, 446–451. Corrigendum (1982), J. Amer. Statist. Assoc. 77, 954.

Kronmal, R.A. and Peterson, A.V. (1984), *An acceptance–complement analogue of the mixture–plus–acceptance–rejection method for generating random variables*, ACM Trans. Math. Software 10, 271–281.

Marsaglia, G. (1968), *Random numbers fall mainly in the planes*, Proc. Nat. Acad. Sciences 61, 25–28.

Neumann, J. v. (1951), *Various techniques used in connection with random digits. Monte Carlo methods*, National Bureau of Standards AMS 12, 36–38.

Niederreiter, H. (1977), *Pseudo-Random Numbers and optimal Coefficients*, Advances in Math. 26, 99–181.

Niederreiter, H. (1978), *Quasi–Monte–Carlo methods and pseudo–random numbers*, Bull. AMS. 84, 957-1041.

Ripley, B.D. (1987), "Stochastic Simulation," J. Wiley, New York.

Stadlober, E. (1989a), *Binomial random variate generation: A method based on ratio of uniforms*, American Journal of Mathematical Management Sciences (to appear).

Stadlober, E. (1989b), *Sampling from Poisson, binomial and hypergeometric distributions: Ratio of uniforms as a simple and fast alternative*, Mathematisch–Statistische Sektion 303. Forschungsgesellschaft Joanneum, Graz, Austria.

Stefanescu, St. and I. Vaduva (1987), *On computer generation of random vectors by transformations of uniformly distributed vectors*, Computing 39, 141–153.

Tadikamalla, P.R. and Johnson (1981), *A complete guide to gamma variate generation*, American Journal of Mathematical and Management Sciences 1, 213–236.

Vaduva, I.(1985), *Computer generation of random vectors based on the transformation of uniformly distributed vectors*, Proceedings of the 7th Conference on Probability Theory, Brasov 1982, VNU Science Press, Utrecht, 589–598.

Walker, A.J. (1974a), *New fast methods for generating discrete numbers with arbitrary frequency distribution*, Electron. Lett. 10, 127–128.

Walker, A.J. (1974b), *Fast generation of uniformly distributed random numbers with floating point representation*, Electron. Lett. 10, 553–554.

Walker, A.J. (1977), *An efficient method for generating discrete random variables with general distributions*, ACM Trans. Math. Software 3, 253–256.

Zechner, H. (1990), *Erzeugung gammaverteilter Zufallszahlen mit allgemeinen Formparametern*, Mathematisch–Statistische Sektion 311. Forschungsgesellschaft Joanneum, Graz, Austria.

Prof. Dr. Ulrich Dieter, Institute of Statistics
Graz University of Technology, A–8010 Graz, Lessingstraße 27, Austria
Tel.: (0043)316–873 6475, bitnet: statistik@ rech.tu–graz.ada.at

Special methods for pseudorandom number generation

Jürgen Lehn

In most applications of stochastic simulation the source of randomness is a sequence of standard pseudorandom numbers $u_0, u_1, u_2, \ldots$, i.e. a sequence of numbers generated by a computer which "behave" as a realization of a sequence $U_0, U_1, U_2, \ldots$ of independent identically distributed random variables having a uniform distribution on the unit interval $[0, 1]$. Many users of simulation do not think about how such numbers are produced by their computers and apply the standard software at hand. This attitude is dangerous since the source of randomness is fundamental for all stochastic simulations and many pseudorandom number generators in use have serious defects. It is also evident that the results of simulation studies depend on the methods of generation of the pseudorandom numbers. The present paper gives a survey of methods for generating deterministic sequences of numbers which can be used as standard pseudorandom number sequences with emphasis on the generation methods studied by the author's research group in Darmstadt.

1 The linear congruential generator

The most frequently used method for the generation of standard pseudorandom numbers goes back to Lehmer (1951). A linear congruential generator is defined by

$$x_i \equiv a \, x_{i-1} + b \pmod{m}$$

for a multiplier a, a shift b, a seed x_0, and a modulus m, where x_0, a, and b are integers between 0 and $m - 1$. If $b = 0$ then the generator is called multiplicative congruential. If all these parameters are chosen carefully, then the sequence of numbers $u_i = x_i/m$ can be taken as a sequence of standard pseudorandom numbers.

The parameters are well chosen if

(i) the maximum period length is achieved,

(ii) the sequence of points $(x_0, \ldots, x_{k-1}), (x_k, \ldots, x_{2k-1}), \ldots$ is not contained in too few hyperplanes.

Conditions which guarantee maximal period length are given e.g. in Knuth (1969). It is well known that the points $(x_0, x_1, \ldots, x_{k-1}), (x_1, x_2, \ldots, x_k), \ldots$ lie on a shifted lattice in the unit cube $[0, 1)^k$, i.e. on a finite number of hyperplanes (see e.g. Marsaglia (1968)). A generator is said to pass the so-called spectral test (see Coveyou and Mc Pherson (1967)) if this number is "large" (see Afflerbach (1990), Ripley (1987), and Knuth (1969)). A basis of the lattice is given by

$$\vec{c_1} = (1, a, a^2, \ldots, a^{k-1}), \vec{c_2} = (0, m, 0, \ldots, 0), \ldots, \vec{c_k} = (0, 0, 0, \ldots, m).$$

This lattice structure of linear congruential pseudorandom numbers has been studied extensively (see e.g. the survey of Niederreiter (1978) or Chapter 3 of Knuth (1969)). The value of the modulus m is often chosen in consideration of the word length of the computer. Since the lattice structure is essentially determined by the multiplier a and the modulus m (the shift parameter b has only influence on the distance and the direction the lattice with basis $\vec{c_1}, \vec{c_2}, \ldots, \vec{c_k}$ is shifted, i.e. its influence is negligible), the problem can be reduced to selecting "good" multipliers a for a given modulus m. Fishman and Moore (1986) carried out an exhaustive search for "good" multipliers a in case of the prime modulus $m = 2^{31} - 1$. Another exhaustive search with the modulus $m = 2^{32}$ and a partial search with $m = 2^{48}$ are reported in Fishman (1990).

A criterion for selecting "good" multipliers can be based on the maximum distance between the hyperplanes carrying the points $(x_0, x_1, \ldots, x_{k-1}), (x_1, x_2, \ldots, x_k), \ldots$ determined by the generator. The reciprocal of this maximum distance can be calculated by use of dual lattices (see Dieter (1975) and Dieter (1986)).

Another way of analyzing the lattice structure of the set of vectors of k consecutive pseudorandom numbers is to consider a Minkowski reduced basis

$$\vec{e_1}, \vec{e_2}, \ldots, \vec{e_k}$$

(which is in some sense a "minimal basis" or a basis of shortest vectors). For the calculation of such a basis from the lattice basis $\vec{c_1}, \vec{c_2}, \ldots, \vec{c_k}$ mentioned above an algorithm of Afflerbach and Grothe (1985) can be applied. The Beyer ratios are defined by

$$q_k = \frac{|\vec{e_1}|}{|\vec{e_k}|}, \quad k = 2, 3, \ldots \quad .$$

Since $|\vec{e_1}| \leq |\vec{e_2}| \leq \ldots \leq |\vec{e_k}|$ follows from the definition of a Minkowski reduced basis, we have $0 < q_k \leq 1$ and generators for which the values of $q_k, k = 2, 3, \ldots$ (up to 20, say), are close to one are sought for (see Beyer, Roof, and Williamson (1971)).

The search for "good" multipliers can also be based on the upper and lower bounds for the k-dimensional discrepancy which have been established by Niederreiter (1985). But there is a lot of other papers dealing with this problem of selecting "good" multipliers for linear congruential generators if the modulus m is given, see e.g. Afflerbach (1990) and the references listed there. Additional work is mentioned in a forthcoming survey paper of Niederreiter (1991a).

2 The matrix generator

A generalization of the linear congruential generator is the multirecursive method defined by

$$x_i \equiv a_1 x_{i-1} + \ldots + a_r x_{i-r} \pmod{p}$$

for a recursion depth r, a prime modulus p, multipliers $a_1, \ldots a_r$, and starting values $x_0, \ldots, x_{r-1}$, all integers between 0 and $p-1$, and not all $x_0, \ldots x_{r-1}$ are equal to zero. These generators were introduced by Knuth (1969) and analized in a Ph.D. thesis (see Grube (1973)) written under the supervision of U. Dieter. Maximum period length $p^r - 1$ is attained, if the multipliers $a_1, \ldots, a_r$ are chosen such that the polynomial

$$x^r - a_1 x^{r-1} - \ldots - a_{r-1} x - a_r$$

is a primitive polynomial mod p, i.e. if this polynomial has a root that is a primitive element of the field with p^k elements. The matrix generator generalizes the multirecursive linear congruential generator. It is defined by

$$\vec{x}_i \equiv A \vec{x}_{i-1} \pmod{p}$$

where $\vec{x}_i$ is an r–vector with integer components, A an $r \times r$ matrix with integer entries, all integers between 0 and $p-1$, and a seed $\vec{x}_0 \neq 0$. Maximum period length $p^r - 1$ is attained by the sequence of r-vectors, i.e. $(p^r - 1) \cdot r$ pseudorandom numbers are generated per period, if the characteristic polynomial of the matrix A is a primitive polynomial mod p. This generation method goes back to a Ph. D. thesis (see Tahmi (1981)) written at the University of Algers and to two papers of Grothe (1986) and of Niederreiter (1986). In the thesis of Tahmi special emphasis is given to the algebraic structure of "degenerate" sequences which do not attain maximum period length. The Ph. D. thesis of Grothe (1988) deals only with matrix generators having maximum period length and contains (up to dimension $r = 6$) proposals for such matrix generators. They have been selected according to their "stochastic quality" which has been judged by considering the lattice structure in higher dimensions.

3 The quadratic congruential generator

Since all these linear generation methods described above create strong linear patterns in all dimensions another approach has been considered, in an attempt to get points "more randomly" spread out. The linear congruential generator can be generalized to a quadratic congruential generator which is defined by the recursion formula

$$x_i \equiv a\, x_{i-1}^2 + b\, x_{i-1} + c \pmod{m}$$

Knuth (1969) proposed this generator and gave conditions on $a, b,$ and c such that the generated sequence has maximum period length m. In Eichenauer and Lehn (1987) it is shown that for such generators with maximum period length the corresponding patterns are superpositions of a certain number of shifted lattices in all dimensions.

4 The inversive congruential generator with prime modulus

In another attempt to get rid of the linear patterns the inversive congruential method has been introduced in Eichenauer and Lehn (1986). The generator is given by the recursion formula

$$x_i \equiv a\, \bar{x}_{i-1} + b \pmod{p}$$

where p is a prime modulus, the seed x_0 as well as the coefficients a and b are integers between 0 and $p-1$, $\bar{x}$ denotes the inverse of x in the multiplicative group of the finite field $GF(p)$ of order p, if $x \neq 0$, and $\bar{0} = 0$. If the polynomial $x^2 - bx - a$ is a primitive polynomial over $GF(p)$ then the generated sequence has maximum period length p. The patterns generated by these generators in higher dimensions k are so irregular that they pass the k–dimensional lattice test for all $k \leq (p+1)/2$, i.e. the coarsest lattice covering all the generated points is the full integer lattice (see Niederreiter (1988) and Eichenauer, Grothe, and Lehn (1988)). This test has been proposed by Marsaglia (1972).

Another interesting result on the "stochastic quality" of the generated sequences has been proved in Niederreiter (1989). Let $u_i = x_i/p$ and $\vec{u}_i = (u_i, \ldots, u_{i+k-1}), i = 0, \ldots, p-1$, correspond to the generated points in the k-dimensional hypercube $[0, 1]^k$. Then the k–dimensional discrepancy is defined by

$$D_p^{(k)} = \sup_J |F(J) - V(J)|$$

where the supremum is taken over all subintervalls J of $[0, 1]^k$ with one vertex at the origin, $F(J)$ is the relative frequency of points falling into J, and $V(J)$ denotes the k–dimensional volume of J. Niederreiter (1990) derived the asymptotic formula

$$D_p^{(k)} = O(p^{-1/2}(\log p)^k)$$

for $2 \leq k < p$ and improved this result by showing that this bound is in the following sense best possible: It holds that for a positive proportion of the admissible parameters (i.e. those which guarantee maximal period length) the discrepancy $D_p^{(k)}$ is at least of the order of magnitude $p^{-1/2}$ for all dimensions $k \geq 2$, in accordance with the idea of points being randomly thrown into the hypercube $[0, 1]^k$.

But the most remarkable result about these generators which supports the hypothesis of having the right degree of irregularity in the generated patterns is contained in a forthcoming paper of Eichenauer–Herrmann (1991) entitled "Inversive congruential pseudorandom numbers avoid the planes". There it is proved that for all dimensions k, $2 \leq k < p$, any hyperplane contains at most k points of the generated k–dimensional pattern. This property of the inversive generators makes them interesting for simulations in Stochastic Geometry.

5 Modifications of the inversive congruential generator

There is also a modification of the inversion method which has been suggested by D.E. Knuth. Instead of a prime modulus p a power of two modulus 2^α can be used. Let $Z = \{0, 1, \ldots, 2^\alpha - 1\}$ and $Z^* = \{1, 3, \ldots, 2^\alpha - 1\}$ for $\alpha \in \mathbb{N}$. Assume $a, b \in Z$, $a \equiv 1 \pmod 4$ and $b \equiv 2 \pmod 4$. Define a nonlinear generator with power of two modulus 2^α by

$$x_i \equiv a\,\bar{x}_{i-1} + b \pmod{2^\alpha}$$

where $\bar{x}$ denotes the uniquely determined number $\bar{x} \in Z^*$ satisfying $x \cdot \bar{x} \equiv 1 \pmod{2^\alpha}$ for $x \in Z^*$ and choose $x_0 \in Z^*$. Then the nonlinear generator has maximum period length $2^{\alpha-1}$ (see Eichenauer, Lehn, and Topuzoğlu 1988). This method and the result on the period length has been generalized recently in Eichenauer–Herrmann and Topuzoğlu (1990) to the case of a general prime power modulus $m = p^\alpha$. The asymptotic behaviour of the two–dimensional discrepancy has been investigated in Niederreiter (1989). The asymptotic formula

$$D_{m/2}^{(2)} = O(m^{-1/2}(\log m)^2)$$

is proved there and in Eichenauer–Herrmann and Niederreiter (1990) it is shown that (as in the prime modulus case) for a positive proportion of the inversive congruential generators with modulus $m = 2^\alpha$ having maximum period length the discrepancy $D_{m/2}^{(k)}$ is at least of the order of magnitude $m^{-1/2}$ for all $k \geq 2$. A precise description of the lattice structure generated by the power of two modulus inversive generators is given in Eichenauer–Herrmann, Grothe, Niederreiter, and Topuzoğlu (1990).

A more detailed discussion of nonlinear congruential generators is contained in a forthcoming paper of Niederreiter (1991b).

References

Afflerbach, L. (1990) On the assessment of linear congruential generators, J.Comp.Appl.Math. 31, 3–10

Afflerbach L. and Grothe, H. (1985) Calculation of Minkowski reduced lattice bases, Computing 35, 269-276

Beyer, W.A., Roof, R.B., and Williamson, D. (1971) The lattice structure of multiplicative congruential pseudo–random vectors, Math.Comp. 25, 345–363

Coveyou, R.R. and Mac Pherson, R.D. (1967) Fourier analysis of uniform random number generators, J.Assoc.Comput.Mach. 14, 100–119

Dieter, U. (1975) How to calculate shortest vectors in a lattice, Math.Comp. 29, 827–833

Dieter, U. (1986) Probleme bei der Erzeugung gleichverteilter Zufallszahlen, in: Afflerbach, L. and Lehn, J. (eds.): "Zufallszahlen und Simulationen", Teubner, Stuttgart, 7-20

Eichenauer, J., Grothe, H., and Lehn, J. (1988) Marsaglia's lattice test and non–linear congruential pseudo random number generators, Metrika 35, 241–250

Eichenauer, J. and Lehn, J. (1986) A non–linear congruential pseudo random number generator, Statist.Papers 27, 315–326

Eichenauer, J. and Lehn, J. (1987) On the structure of quadratic congruential sequences, manuscripta math. 58, 129–140

Eichenauer, J., Lehn, J., and Topuzoğlu, A. (1988) A nonlinear congruential pseudorandom number generator with power of two modulus, Math.Comp. 51, 757–759

Eichenauer–Herrmann, J. (1991) Inversive congruential pseudorandom numbers avoid the planes, Math.Comp. (to appear)

Eichenauer–Herrmann, J., Grothe, H., Niederreiter, H., and Topuzoğlu, A. (1990) On the lattice structure of a nonlinear generator with modulus 2^α, J. Comp. Appl. Math. 31, 81–85

Eichenauer–Herrmann, J. and Niederreiter, H. (1990) Lower bounds for the discrepancy of inversive congruential pseudorandom numbers with power of two modulus, Fachbereich Mathematik, Preprint Nr. 1302, Darmstadt

Eichenauer–Herrmann,J. and Topuzoğlu, A. (1990) On the period length of congruential pseudorandom number sequences generated by inversions, J.Comp.Appl.Math. 31, 87–96

Fishman, G.S. and Moore, L.R. (1986) An exhaustive analysis of multiplicative congruential random number generators with modulus $2^{31} - 1$, SIAM J.Sci.Statist. Comp.7, 24–45; Erratum, ibid. 7, 1058

Fishman, G.S. (1990) Multiplicative congruential random number generators with modulus 2^{β}: An exhaustive analysis for $\beta = 32$ and a partial analysis for $\beta = 48$, Math.Comp. 54, 331-344

Grothe, H. (1986) Matrix–Generatoren, in: Afflerbach, L., Lehn, J. (eds.): "Zufallszahlen und Simulationen", Teubner, Stuttgart, 29–34

Grothe, H. (1988) Matrix–Generatoren zur Erzeugung gleichverteilter Pseudozufallsvektoren, Thesis, Darmstadt

Grube, A. (1973) Mehrfach rekursiv erzeugte Zufallszahlen, Thesis, Karlsruhe

Knuth, D.E. (1969) The Art of Computer Programming, Vol. 2, 1st ed., Addison–Wesley, Reading (Mass.)

Lehmer, D.H. (1951) Mathematical methods in large–scale computing units, Proc. 2nd Symposium on Large–Scale Digital Calculating Machinery, 141–146, Harvard Univ.Press, Cambridge, Mass.

Marsaglia, G. (1968) Random numbers fall mainly in the planes, Proc.Nat.Acad.Sci. USA 61, 25–28

Marsaglia, G. (1972) The lattice structure of linear congruential sequences, in: Zaremba, S.K. (ed.): "Applications of Number Theory to Numerical Analysis", Academic Press, New York (1972), 249–285

Niederreiter, H. (1978) Quasi-Monte Carlo methods and pseudo–random numbers, Bull.Amer. Math. Soc. 84, 957–1041

Niederreiter, H. (1985) The serial test for pseudo–random numbers generated by the linear congruential method, Numer.Math. 46, 51–68

Niederreiter, H. (1986) A pseudorandom vector generator based on finite field arithmetic, Math. Japonica 31, 759–774

Niederreiter, H. (1988) Remarks on nonlinear congruential pseudorandom numbers, Metrika 35, 321–328

Niederreiter, H. (1989) The serial test for congruential pseudorandom numbers generated by inversions, Math.Comp. 52, 135–144

Niederreiter, H. (1990) Lower bounds for the discrepancy of inversive congruential pseudorandom numbers, Math.Comp. 55, 277–287

Niederreiter, H. (1991a) Recent trends in random number and random vector generation, (Proc. 5th Internat.Conf. on Stochastic Programming, Ann Arbor, 1989), Ann. Operations Research (to appear)

Niederreiter, H. (1991b) Nonlinear methods for pseudorandom number and vector generation. Proc.Internat.Workshop on Computationally-Intensive Methods in Simulation and Stochastic Optimization, Laxenburg 1990 (to appear)

Ripley, B.D. (1987) Stochastic Simulation, John Wiley & Sons, New York

Tahmi, E.–H. (1982) Contribution aux générateurs des vecteurs pseudo–aléatoires, Thesis, Algers

Jürgen Lehn
Department of Mathematics
Technische Hochschule Darmstadt
Schlossgartenstr. 7
D – 6100 Darmstadt
Federal Republic of Germany

Monte-Carlo-Techniques in Inferential Statistics (2)

Designing Bootstrap Prediction Regions

Rudolf Beran[*]

University of California, Berkeley

ABSTRACT

This article discusses the design and bootstrap construction of asymptotically optimal prediction regions. The emphasis is on devising simultaneous one-sided prediction intervals. A good solution to this problem implies constructions for simultaneous two-sided prediction intervals and for multivariate prediction regions.

1. INTRODUCTION

Prediction regions will be discussed in the following context. An observed learning sample Y_n and a potentially observable variable X have a joint distribution $P_{\theta,n}$. The unknown parameter θ lies in a parameter space Θ. To be predicted are the potentially observable variables $\{Z_u = Z_u(X) : u \in U\}$, a specified collection of functions of X labelled by an index set U. In this article, Θ is an open subset of a euclidean space and U is a finite set with elements $\{u_1, u_2, \ldots, u_r\}$.

The development will emphasize the case where the $\{Z_u : u \in U\}$ are real-valued random variables and the goal is to devise simultaneous one-sided prediction intervals D_n for the $\{Z_u\}$. By appropriate choice of the functions $\{Z_u\}$, the analysis for this one-sided case applies as well to simultaneous two-sided prediction intervals and to multivariate prediction regions. This point will be illustrated through examples.

Formally, let $z = \{z_u : u \in U\}$ denote a generic value of $Z = \{Z_u : u \in U\}$. Define the prediction region

$$D_n = \{z : z_u \leq c_{n,u} \text{ for every } u \in U\}, \tag{1.1}$$

where each of the critical values $c_{n,u} = c_{n,u}(Y_n)$ depends on the learning sample. Evidently, D_n is equivalent to simultaneously asserting the one-sided prediction intervals

$$D_{n,u} = \{z_u : z_u \leq c_{n,u}\} \quad u \in U \tag{1.2}$$

for the respective variables Z_u.

The coverage probability of D_n or if each $D_{n,u}$ can be assessed conditionally, given the learning sample, or unconditionally. Let $P_\theta(\cdot | Y_n)$ denote the conditional distribution of X given Y_n. The *conditional coverage probability* of D_n given Y_n is

$$CP(D_n | Y_n, \theta) = P_\theta(Z \in D_n | Y_n). \tag{1.3}$$

The *unconditional coverage probability* of D_n is

$$CP(D_n | \theta) = P_{\theta,n}(Z \in D_n) \tag{1.4}$$

$$= E_\theta CP(D_n | Y_n, \theta)$$

where the expectation is calculated with respect to the distribution $Q_{\theta,n}$ of Y_n. The notations $CP(D_{n,u} | Y_n, \theta)$ and $CP(D_{n,u} | Y_n, \theta)$ similarly denote the conditional and unconditional coverage probabilities of $D_{n,u}$. The conditional coverage probability concept has been emphasized by Box and Jenkins (1976) in prediction of time series and by writers on tolerance regions (cf. Guttman 1970). The unconditional coverage probability concept has been supported by Cox and Hinkley (1974), among others. For additional background references on prediction regions, see Beran (1990a) and Carroll and Ruppert (1989).

---

[*] This research was supported in part by NSF Grant DMS 9001710.

How should we choose the critical values $\{c_{n,u} : u \in U\}$ in the definition (1.1) of D_n? Suppose the real random variables $\{Z_u : u \in U\}$ measure logically similar attributes. In addition, suppose that X and Y_n are independent. It is natural then to impose two design requirements on the critical values $\{c_{n,u}\}$:

Design Goal 1. Choose the $\{c_{n,u}\}$ so that

$$CP(D_n \mid Y_n, \theta) \rightarrow \alpha \text{ in } Q_{\theta,n} \text{ probability} \tag{1.5}$$

as n increases. The value of the constant α in $(0, 1)$ is pre-specified.

Design Goal 2. Having met design goal 1, choose the $\{c_{n,u}\}$ so that, for every u in U,

$$CP(D_{n,u} \mid Y_n, \theta) \rightarrow \beta(\alpha, \theta) \text{ in } Q_{\theta,n} \text{ probability} \tag{1.6}$$

for some constant $\beta(\alpha, \theta)$ which does not depend on u but can depend on α and θ.

The first design goal controls the asymptotic simultaneous coverage probability of D_n, both conditionally and unconditionally. The second design goal controls the *relative* asymptotic conditional and unconditional coverage probabilities of the $\{D_{n,u}\}$. Indeed, since the $\{Z_u\}$ are assumed logically similar, it is only fair to require that the $\{D_{n,u}\}$ have equal coverage probabilities. Design goal 2 has been called conditional asymptotic *balance* of the simultaneous prediction region D_n (Beran 1988 and 1990b). When X and Y_n are dependent, the formulation of goal 2 becomes more complex technically.

In many examples, more than one construction of the critical values $\{c_{n,u}\}$ will satisfy design goals 1 and 2. Among these possibilities, how might the choice be narrowed further? Considering the conditional and unconditional errors in the balance and in the simultaneous coverage probability of D_n suggests two additional design goals:

Design Goal 3. Having met design goals 1 and 2, choose the $\{c_{n,u}\}$ so as to minimize asymptotically the dispersions of $CP(D_n \mid Y_n, \theta) - \alpha$ and of the $\{CP(D_{n,u} \mid Y_n, \theta) - \beta(\alpha, \theta)\}$

Design Goal 4. Having met design goals and and 2, choose the $\{c_{n,u}\}$ so as to maximize the rates of convergence to 0 of $CP(D_n \mid \theta) - \alpha$ and of the $\{CP(D_{n,u} \mid \theta) - \beta(\alpha, \theta)\}$ as n increases.

Goal 4 addresses asymptotic biases in the conditional simultaneous coverage probability and balance of D_n while goal 3 seeks to control the dispersions of the various conditional coverage probabilities. For the models treated in this article, design goals 3 and 4 are mutually compatible.

2. DESIGN GOALS 1 AND 2

As in the latter part of the Introduction, suppose X and Y_n are independent, with distributions P_θ and $Q_{\theta,n}$ respectively. Two further sets of conditions are used in the analysis of this section:

Assumption A. The cdf's

$$A_u(x, \theta) = P_\theta(Z_u \leq x) \quad u \in U \tag{2.1}$$

are continuous and strictly monotone in x for every θ.

Assumption B. The cdf

$$B(x_1, \ldots, x_r, \theta) = P_\theta(Z_{u_1} \leq x_1, \ldots, Z_{u_r} \leq x_r) \tag{2.2}$$

is continuous in $(x_1, \ldots, x_r)$ for every θ. The cdf

$$A(x, \theta) = B[A_{u_1}^{-1}(x, \theta), \ldots, A_{u_r}^{-1}(x, \theta), \theta] \tag{2.3}$$

is continuous and strictly monotone for every θ.

These assumptions yield a convenient necessary and sufficient condition for design goals 1 and 2.

Proposition 1. Suppose Assumptions A and B hold. Then, design goals 1 and 2 are met if and only if

$$c_{n,u} \to A_u^{-1}[A^{-1}(\alpha, \theta), \theta] \text{ in } Q_{\theta,n} \text{ probability} \tag{2.4}$$

for every u. In that event, the limit in (1.6) is

$$\beta(\alpha, \theta) = A^{-1}(\alpha, \theta) \tag{2.5}$$

Remark A. The proof of this result rests on the expressions

$$CP(D_{n,u}|Y_n, \theta) = A_u(c_{n,u}, \theta) \tag{2.6}$$

$$CP(D_n|Y_n, \theta) = B(c_{n,u_1}, \ldots, c_{n,u_r}, \theta).$$

For details, see Beran (1990c).

Remark B. Suppose $\hat{\theta}_n = \hat{\theta}_n(Y_n)$ is an estimate of θ such that $\hat{\theta}_n \to \theta$ in $Q_{\theta,n}$ probability. Then, the choice of critical values

$$c_{n,u} = A_u^{-1}[A^{-1}(\alpha, \hat{\theta}_n), \hat{\theta}_n]$$

meets design goals 1 and 2 under the conditions of Proposition 1.

Remark C. A bootstrap algorithm can be used to approximate the cdf's $\{A_u(\cdot, \theta)\}$ and $A(\cdot, \theta)$. The basis for this algorithm is the definition (2.1) and the expression

$$A(x, \theta) = P_\theta[\sup_u A_u(Z_u, \theta) \le x]. \tag{2.7}$$

For related algorithms, see Beran (1988) and (1990a).

Example 1. Let the $\{X_i : i \ge 1\}$ be a sequence of i.i.d. $N(\mu, \sigma^2)$ random variables, the parameter $\theta = (\mu, \sigma^2)$ being unknown. On the basis of the learning sample $Y_n = (X_1, \ldots, X_n)$, the aim is to predict simultaneously the r components $\{Z_u\}$ of $X = \{X_{n+1}, \ldots, X_{n+r}\}$ for given positive r. Let $\hat{\theta}_n = (\overline{X}_n, s_n^2)$, let Φ denote the standard normal cdf, and let $w(\alpha, r) = \Phi^{-1}(\alpha^{1/r})$. Since

$$A_u(x, \theta) = \Phi[\sigma^{-1}(x - \mu)] \quad 1 \le u \le r \tag{2.8}$$

$$A(x, \theta) = x^r \quad 0 \le x \le 1$$

it follows from Proposition 1 and Remark B that the simultaneous one-sided prediction intervals

$$D_n = (-\infty, \overline{X}_n + s_n w(\alpha, r)]^r \tag{2.9}$$

for $X_{n+1}, \ldots, X_{n+r}$ satisfy design goals 1 and 2.

To construct simultaneous two-sided prediction intervals in this example, set $Z_u = X_{n+u}$ and $Z_{u+r} = -X_{n+u}$ for $1 \le u \le r$. Simultaneous one-sided prediction intervals for these $\{Z_u : 1 \le u \le 2r\}$ are equivalent to simultaneous two-sided prediction intervals for the variables $(X_{n+1}, \ldots, X_{n+r})$ of interest. Now

$$A_u(x, \theta) = \Phi[\sigma^{-1}(x - \mu)] \quad 1 \le u \le r$$

$$A_u(x, \theta) = \Phi[\sigma^{-1}(x + \mu)] \quad r+1 \le u \le 2r \tag{2.10}$$

$$A(x, \theta) = (2x - 1)^r \quad 1/2 \le x \le 1.$$

By Proposition 1 and Remark B, the simultaneous two-sided prediction intervals

$$D_n = [\overline{X}_n - s_n w\{(1 + \alpha)/2, r\}, \quad \overline{X}_n + s_n w\{(1 + \alpha)/2, r\}]^r \tag{2.11}$$

for $X_{n+1}, \ldots, X_{n+r}$ satisfy design goals 1 and 2. In the two-sided case, design goal 2 can be re-interpreted as asymptotic centering plus asymptotic balance of the constituent two-sided intervals that make up D_n.

Example 2. Let the $\{X_i : i \ge 1\}$ be iid r-variate random vectors with unknown cdf F which has full support and is absolutely continuous. On the basis of learning sample $Y_n = (X_1, \ldots, X_n)$, the aim is to predict simultaneously the r components of $X = X_{n+1}$. Here θ is F, a function-valued parameter. Nevertheless, Proposition 1 and Remark B can be extended to this setting (see Beran 1990c) as follows.

Take U to be the standard basis vectors for R^r and their negatives. Define $Z_u = u'X$ for every vector u in U. Let $J_u(x, F)$ denote the cdf of Z_u and let $J(x, F)$ denote the cdf of $\max\{J_u(Z_u, F) : u \in U\}$. Finally, let $\hat{F}_n$ be the empirical cdf of the learning sample. Then, the critical values

$$c_{n,u} = J_u^{-1}[J^{-1}(\alpha, \hat{F}_n), \hat{F}_n], \quad u \in U \tag{2.12}$$

and (1.1) yield simultaneous two-sided prediction intervals for the components of X that satisfy both design goals 1 and 2.

Mardia, Kent and Bibby (1979) reported test scores for 88 college students, each of whom wrote two-closed book and three open-book tests. Taking the closed-book scores alone as learning sample (n = 88, r = 2) and $\alpha = .90$, the preceding paragraph generates the prediction box $D_n = [7, 64] \times [27, 72]$ for the scores of a future student taking the same closed-book tests. The balancing of D_n is in the four axial directions given by U. For this D_n, the quantity $\beta(\alpha, F)$ in design goal 2 is estimated by $\beta(\alpha, \hat{F}_n) = 85/88 = .966$. To see this, plot D_n on a scatterplot of the learning sample.

3. DESIGN GOAL 3

Having satisfied design goals 1 and 2, we seek to control the dispersion of the conditional coverage probabilities of D_n and the $\{D_{n,u}\}$. Formally, let $\{D_n(d) : n \ge 1\}$ be a sequence of simultaneous prediction intervals for $(Z_{u_1}, \ldots, Z_{u_n})$ of the form (1.1), where $d = \{(c_{n,u_1}, \ldots, c_{n,u_n}) : n \ge 1\}$ denotes any sequence of critical values. Write

$$D_{n,u}(d) = \{z_u : z_u \le c_{n,u}\} \tag{3.1}$$

for the uth component interval in $D_n(d)$.

Suppose the sequence d satisfies (2.4), so that goals 1 and 2 are met. Let

$$T_n(d, \theta) = n^{1/2}\{CP[D_n(d)|Y_n, \theta] - \alpha\} \tag{3.2}$$

$$T_{n,u}(d, \theta) = n^{1/2}\{CP[D_{n,u}(d)|Y_n, \theta] - A^{-1}(\alpha, \theta)\}.$$

Let w be a monotone increasing function on the non-negative reals, with $w(0) = 0$. Measure the dispersions of the conditional coverage probabilities through the risks

$$\rho_n(d, \theta) = E_\theta w[|T_n(d, \theta)|] \tag{3.3}$$

$$\rho_{n,u}(d, \theta) = E_\theta w[|T_{n,u}(d, \theta)|].$$

We obtain local asymptotic minimax lower bounds on these risks, using the following classical regularity assumption on the model $Q_{\theta,n}$. For notational simplicity, we take θ to be real-valued.

Assumption C. For $\theta_n = \theta + n^{-1/2}h$, where h is real, let $Q_{\theta_n,n}^c$ and $Q_{\theta_n,n}^s$ denote, respectively, the absolutely continuous and singular parts of $Q_{\theta_n,n}$ with respect to $Q_{\theta,n}$. Let $L_n(h,\theta)$ denote the log-likelihood ratio of $Q_{\theta_n,n}^c$ with respect to $Q_{\theta,n}$. There exist random variables $\xi_n(\theta)$, depending on Y_n and on θ, and a positive constant $I(\theta)$ such that

$$L_n(h,\theta) - h\xi_n(\theta) + 2^{-1}h^2 I(\theta) \to 0 \qquad (3.4)$$

in $Q_{\theta,n}$ probability, for every real h, and

$$L[\xi_n(\theta)|\theta] \Rightarrow N(0,I(\theta)). \qquad (3.5)$$

Without any loss of generality, we can assume that $\xi_n(\theta)$ is constructed so that

$$\xi_n(\theta_n) - \xi_n(\theta) + h I(\theta) \to 0 \qquad (3.6)$$

in $Q_{\theta,n}$ probability for every real h (Le Cam 1969, p.68).

Define, on the unit interval, the cdf's

$$\tilde{A}(x,\theta,t) = B[A_{u_1}^{-1}(x,t),\ldots,A_{u_r}^{-1}(x,t),\theta] \qquad (3.7)$$

and

$$C_u(x,\theta,t) = A_u[A^{-1}\{A^{-1}(x,t),t\},\theta] \qquad (3.8)$$

$$C(x,\theta,t) = \tilde{A}[A^{-1}(x,t),\theta,t].$$

Note that $C_u(x,\theta,\theta) = A^{-1}(x,\theta)$ and $C(x,\theta,\theta) = x$. In the sequel, notation like $C^{(i,j,k)}(x,\theta,t)$ will represent the partial derivative $\partial^{i+j+k} C(x,\theta,t)/\partial x^i \partial\theta^j \partial t^k$.

Proposition 2. Suppose Assumption C holds and the sequences d considered all satisfy (2.4). Suppose that $A^{(1,0)}(x,\theta)$ and $\{C_u^{(0,0,1)}(x,\theta,t)\}$ exist and are continuous in (x,θ,t) at all points where $t = \theta$. Then

$$\lim_{b\to\infty} \liminf_{n\to\infty} \inf_d \sup_{|h|\le b} \rho_{n,u}(d,\theta) \ge Ew[|\tau_u(\theta)W|] \qquad (3.9)$$

where W is a standard normal random variable and

$$\tau_u^2(\theta) = [C_u^{(0,0,1)}(\alpha,\theta,\theta)]^2 I^{-1}(\theta). \qquad (3.10)$$

Suppose that $\{\partial B(x_1,\ldots,x_n,\theta)/\partial x_i : 1 \le i \le r\}$ and $C^{(0,0,1)}(x,\theta,t)$ exist and are continuous in (x,θ,t) at all points where $t = \theta$. Then the risk $\rho_n(d,\theta)$ satisfies a lower bound like (3.9) in which $\tau_u^2(\theta)$ is replaced by

$$\tau^2(\theta) = [C^{(0,0,1)}(\alpha,\theta,\theta)]^2 I^{-1}(\theta) \qquad (3.11)$$

Remark D. Suppose $\hat{\theta}_n$ and the critical values $\{c_{n,u}\}$ are chosen so that

$$n^{1/2}(\hat{\theta}_n - \theta) - I^{-1}(\theta)\xi_n(\theta) \to 0 \qquad (3.12)$$

$$n^{1/2}\{c_{n,u} - A_u^{-1}[A^{-1}(\alpha,\hat{\theta}_n),\hat{\theta}_n]\} \to 0$$

in $Q_{\theta,n}$ probability, for every u in U. Then, under the conditions of Proposition 2, the two lower bounds on risk are attained, in the sense that

$$\lim_{n\to\infty} \sup_{|h|\le b} \rho_{n,u}(d,\theta) = Ew[|\tau_u(\theta)W] \qquad (3.13)$$

$$\lim_{\substack{n\to\infty \\ |h|\le b}} \sup \rho_n(d,\theta) = Ew[|\tau_u(\theta)W|]$$

for every positive b. Proofs of the Remark and Proposition 2 follow from arguments in Beran (199?).

Remark E. In view of Remarks B and D, the simplest choice of critical values $\{c_{n,u}\}$ that meets design goals 1,2 and 3 is

$$c_{n,u} = A_u^{-1}[A^{-1}(\alpha,\hat{\theta}_n),\hat{\theta}_n] \tag{3.14}$$

with $\hat{\theta}_n$ an asymptotically efficient estimate of θ. Examples 1 and 2 illustrate precisely this construction of the critical values for D_n.

Remark F. Hájek-style convolution representations can be proved for the limiting distributions of $T_n(d,\theta)$ and $\{T_{n,u}(d,\theta)\}$. This approach provides another way of justifying the conditions (3.12) for minimizing the asymptotic dispersions of $T_n(d,\theta)$ and $\{T_{n,u}(d,\theta)\}$,

4. DESIGN GOAL 4

Having satisfied design goals 1,2 and 3 through construction (3.14) for the critical values, it remains to consider design goal 4 — the rates at which the unconditional coverage probabilities of D_n and of the $\{D_{n,u}\}$ converge to α and to $A^{-1}(\alpha,\theta)$ respectively. The key to the discussion is the following assumption on $\hat{\theta}_n$:

Assumption D. For $j = 1,2$, there exist functions $a_j(\theta)$ such that

$$E_\theta(\hat{\theta}_n - \theta)^j = n^{-1}a_j(\theta) + O(n^{-2}) \tag{4.1}$$

as n increases.

This assumption is satisfied, for instance, when $\{Q_{\theta,n}\}$ is a smoothly parametrized exponential family and $\hat{\theta}_n$ is the maximum likelihood estimate of θ.

Proposition 3. Suppose Assumptions A, B and D hold and the critical values $\{c_{n,u}\}$ are given by (3.14). Suppose $C^{(0,0,2)}(x,\theta,t)$ and $\{C_u^{(0,0,2)}(x,\theta,t)\}$ exist and are continuous in (θ,t) at all points where $t = \theta$. Then

$$CP(D_n|\theta) = \alpha + n^{-1}b_1(\alpha,\theta) + O(n^{-2}) \tag{4.2}$$

$$CP(D_{n,u}|\theta) = A^{-1}(\alpha,\theta) + n^{-1}b_{1,u}(\alpha,\theta) + O(n^{-2})$$

where

$$b_1(\alpha,\theta) = \sum_{j=1}^{2} a_j(\theta)C^{(0,0,j)}(\alpha,\theta,\theta)/j! \tag{4.3}$$

$$b_{1,u}(\alpha,\theta) = \sum_{j=1}^{2} a_j(\theta)C_u^{(0,0,j)}(\alpha,\theta,\theta)/j!$$

Remark G. This result follows from Proposition 1A in Beran (1990b). Proposition 3 asserts that, when the critical values for D_n are given by (3.14), the errors in the unconditional balance and overall coverage probability of D_n will typically be of order $O(n^{-1})$. On the other hand, the second line in (3.12) implies that the $\{c_{n,u}\}$ may be perturbed by $o_P(n^{-1/2})$ without affecting the fulfillment of design goals 1,2 and 3. This possibility is the key to the success of the calibration perturbation described below.

Let

$$S_{n,u} = A[A_u(Z_u, \hat{\theta}_n), \hat{\theta}_n] \tag{4.4}$$

and define the cdf's

$$H_{n,u}(x, \theta) = P_{\theta,n}[S_{n,u} \leq x] \tag{4.5}$$

$$H_n(x, \theta) = P_{\theta,n}[\sup_u H_{n,u}(S_{n,u}, \theta) \leq x].$$

Note that $H_{n,u}(\alpha, \theta)$ is just $CP(D_{n,u}|\theta)$ when the critical values are given by (3.14).

Proposition 4. Suppose the critical values $\{c_{n,u}\}$ are given by

$$c_{n,u} = A_u^{-1}[A^{-1}\{H_{n,u}^{-1}[H_n^{-1}(\alpha, \hat{\theta}_n), \hat{\theta}_n], \hat{\theta}_n\}, \hat{\theta}_n]. \tag{4.6}$$

Under Assumption D and regularity conditions described in Proposition 2A of Beran (1990b),

$$CP(D_n|\theta) = \alpha + O(n^{-2}) \tag{4.7}$$

$$CP(D_{n,u}|\theta) = H_n^{-1}(\alpha, \theta) + O(n^{-2}).$$

Remark H. The critical values defined by (4.6) differ from those defined in (3.14) by $O_P(n^{-1})$, under the conditions of Proposition 4. Consequently, for the purposes of design goals 1, 2 and 3, it does not matter whether the $\{c_{n,u}\}$ are given by (3.14) or (4.6). However, the latter critical values are preferable for design goal 4; compare the respective conclusions (4.2) and (4.7) of Proposition 3 and 4.

Remark I. Computing the cdf's $H_n(\cdot, \theta)$ and $\{H_{n,u}(\cdot, \theta)\}$ usually requires another round of bootstrap approximation (cf. Beran 1990b).

Example 1 (continued). Let F_ν denote the cdf of the t-distribution with ν degrees of freedom. For the simultaneous one-sided prediction intervals

$$H_{n,u}(\alpha, \theta) = F_{n-1}[w(\alpha, r)(1 + n^{-1})^{-1/2}]$$

$$H_n(\alpha, \theta) = P_{\theta,n}[\sup_{1 \leq u \leq r} F_{n-1}\{(1 + n^{-1})^{-1/2} s_n^{-1}(Z_u - \overline{X}_n) \leq \alpha] \tag{4.8}$$

$$= J_n(\alpha), \quad \text{say},$$

where J_n is a cdf which does not depend on $\theta = (\mu, \sigma^2)$. The improved critical values (4.6) for this example are thus

$$c_{n,u} = \overline{X}_n + (1 + n^{-1})^{1/2} s_n F_{n-1}^{-1}[J_n^{-1}(\alpha)] \tag{4.9}$$

$$= c_n, \quad \text{say},$$

in contrast to those in (2.9). The corresponding simultaneous one-sided prediction intervals

$$D_n = (-\infty, c_n]^r \tag{4.10}$$

fullfil design goals 1 to 4 ideally. Indeed,

$$CP(D_n|\theta) = \alpha \tag{4.11}$$

$$CP(D_{n,u}|\theta) = J_n^{-1}(\alpha)$$

exactly, a result even better than promised by Proposition 4.

REFERENCES

Beran, R. (1988), "Balanced simultaneous confidence sets," *Journal of the American Statistical Association*, 83, 679-697.

Beran, R. (1990a), "Calibrating prediction regions," *Journal of the American Statistical Association*, 85, 715-723.

Beran, R. (1990b), "Simultaneous prediction regions," unpublished preprint.

Beran, R. (1990c), "Probability-centered prediction regions," unpublished preprint.

Beran, R. (199?), "Controlling conditional coverage probability in prediction," *Ann. Statist.*, to appear.

Box, G.E.P. and Jenkins, G.M. (1976), *Time Series Analysis: Forecasting and Control*, revised edition, Oakland: Holden-Day.

Carroll, R.J. and Ruppert, D. (1989), "Prediction and tolerance intervals with transformation and/or weighting," unpublished preprint.

Cox, D.R. and Hinkley, D.V. (1974), *Theoretical Statistics*, London: Chapman and Hall.

Mardia, K.V., Kent, J.T., and Bibby, J.M. (1979), *Multivariate Analysis*, New York: Academic Press.

THE GENERALIZED BOOTSTRAP

Edward J. Dudewicz

Department of Mathematics, Syracuse University
Syracuse, New York 13244-1150, U.S.A.

ABSTRACT. The bootstrap of Efron is generalized to the Generalized Bootstrap. The Generalized Bootstrap has superior properties for continuous data in the small-sample non-asymptotic case.

INTRODUCTION

The terms "Monte Carlo," "Jackknife," "Randomization Test," "Resampling," "Sample Reuse," "Bootstrap," and "Simulation" are all applied to various methods for dealing with statistical estimation (or confidence, or selection, or other) problems based on a limited set of data with an (at most) partially known distribution function (d.f.).

To set a framework for this paper, let $F(\cdot)$ be a d.f. of a random variable (r.v.) X, and let $\theta(F)$ be some function of F which is of interest.

If $F(\cdot)$ is known, e.g. if $F(x) = \Phi(x)$, the standard normal d.f., and if $\theta(F)$ is $\int \cos^{-1}(x^{10}) dF(x)$, then one may seek to evaluate $\theta(F)$ exactly by (e.g.) integration.

If $F(\cdot)$ is known, but $\theta(F)$ is difficult to evaluate exactly, one may generate independent r.v.'s $X_1, X_2, \ldots, X_n$ with d.f. $F(\cdot)$ and estimate $\theta(F)$ by

$$\frac{g(X_1) + g(X_2) + \ldots + g(X_n)}{n} \tag{1}$$

where in the example $g(X_i) = \cos^{-1}(X_i^{10})$ (i=1,2,...,n), which is called a <u>Monte Carlo Method</u>. It is widely known that a Monte Carlo method was used by W.S. Gosset ("Student") in 1908 to bolster his faith in his t-distribution, and that the term Monte Carlo was introduced by such workers as N. Metropolis and S. Ulam around 1949. The method has been documented in use as early as 1876 by the American statistician E.L. De Forest (see Gentle (1985), p. 612).

1. JACKKNIFE AND BOOTSTRAP

The <u>Jackknife Method</u> was introduced by Quenouille (1949), and named by Tukey in 1958, for reducing bias of an estimator $\hat{\theta}$. One version divides the random sample into g groups of equal size m = n/g observations. It then deletes one group at a time and estimates $\theta(F)$ based on the remaining (g-1)m observations, using the same estimation procedure previously used with a sample of size n. If $\hat{\theta}_i$ is the estimator of θ with group i deleted, we call $\hat{\theta}_i$ a <u>jackknife statistic</u>. The <u>jackknife estimator</u> of θ is

$$J(\hat{\theta}) = \sum_{i=1}^{g} J_i/g, \tag{2}$$

where

$$J_i = g\hat{\theta} - (g-1)\hat{\theta}_i \tag{3}$$

is called a <u>pseudovalue</u>. The idea is to use $J(\hat{\theta})$ in preference to $\hat{\theta}$, and to form a putative confidence interval (CI) for $\theta(F)$ by treating (when g=1)

$$\frac{\sqrt{n} \ (J(\hat{\theta}) - \theta)}{\left(\sum_{i=1}^{n} \frac{(J_i - J(\hat{\theta}))^2}{n-1}\right)^{0.5}} \tag{4}$$

as approximately t_{n-1}, an idea due to Tukey (1958), which has not been verified as yielding a legitimate CI unless n→∞ (see Efron (1982), p. 14).

<u>Randomization Tests</u>, which we mention for historical completeness, compute a test statistic from the data and then for each possible permutation (or division, or re-arrangement) of it. The original test statistic value is evaluated by a p-value with regard to its place in the whole set in determining significance. Such tests were introduced by R.A. Fisher in 1935, and built on by O. Kempthorne for F-test randomization. (For further details, see Edgington (1986).) This procedure and related procedures such as the jackknife are sometimes termed <u>Resampling Procedures</u> (as they can be viewed as recomputing the statistic many times with reweighted sample values) or <u>Sample ReUse Procedures</u> (as one can also take the view that after the sample is used to compute the statistic, it is reused to evaluate or imporve performance).

The <u>Bootstrap Method</u> is almost universally stated as follows (e.g., see Hinkley (1983), Swanepoel (1985), or Csörgö and Mason (1989)). Form the empirical d.f. F_n of the data $X_1, X_2, \ldots, X_n$. Then from N successive random samples with replacement of size n from F_n find estimates $\hat{\theta}_1, \hat{\theta}_2, \ldots, \hat{\theta}_N$ of $\theta(F)$. Then

$$\{\text{Statistic of Interest Calculated from Distribution of } \hat{\theta}_1, \hat{\theta}_2, \ldots, \hat{\theta}_N\} \tag{5}$$

is called the <u>Bootstrap Estimator</u> of $\theta(F)$, and percentiles of the histogram of $\hat{\theta}_1, \hat{\theta}_2, \ldots, \hat{\theta}_N$ are used to form <u>Bootstrap CI's</u>. (Sometimes one may theoretically evaluate the distribution of $\hat{\theta}_i$ under random samples from F_n, instead of using a Monte Carlo estimate, e.g. see Swanepoel (1985), but usually this is not possible.) This method was given its name by Efron in 1979 (see Efron (1982)), and has since seen an explosion of work. Recently whole conferences have been devoted to the topic (e.g., IMS Special Topics Meeting on the Bootstrap, May 15-16, 1990, Michigan State University, with 32 papers; and The International Conference on Bootstrapping and Related Techniques, June 4-8, 1990, University of Trier, West Germany, with 51 papers).

It has been argued in several places by several authors that, while Efron's naming of the procedure seemed to give it a popular impetus, it was in fact in use as early as 1967 (e.g., see Simon (1969a,b), and Dudewicz and Mishra (1988)).

2. THE GENERALIZED BOOTSTRAP GB

We have argued elsewhere (e.g. see discussion of Swanepoel (1985)) that the EDF (empirical d.f.) is a rough estimator of F, and one would do better with a method that smoothed by some density estimation method, then sampled from the smoothed distribution: if you <u>know</u> you have a continuous distribution, you would be ill-advised (except asymptotically, where sample size conquers all) to use the discontinuous F_n. We now provide such a method.

We formulate <u>The Generalized Bootstrap</u> $GB(\theta, F, X_1, \ldots, X_n, F_c, \hat{F}, n, N)$ as:

<u>Step GB-1.</u> The problem is to study the quantity $\theta(F)$, a function of an unknown d.f. $F(\cdot)$.

<u>Step GB-2.</u> A random sample $X_1, \ldots, X_n$ is available from $F(\cdot)$. Based on it, estimate $\theta(F)$ by $\hat{\theta}$.

<u>Step GB-3.</u> It is known that $F \in F_c$, a class of d.f.'s. [If F is known completely, F_c contains one d.f. If F is known to be (e.g.) normal but with unknown mean and/or variance, then F_c is that class of distributions. If F is known to be continuous, then F_c is the class of continuous d.f.'s.]

<u>Step GB-4.</u> Estimate $F \in F_c$ by $\hat{F}$. [One might choose $\hat{F} = F_n$, the empiric d.f. Of course if F_c consists of all continuous d.f.'s, then one will have an estimate $\hat{F}$ such that $\hat{F} \not\in F_c$. Other estimates $\hat{F}$ are discussed below.]

<u>Step GB-5.</u> Independently generate N random samples of size n from $\hat{F}$. From the first sample, say
$$Y_1, Y_2, \ldots, Y_n, \text{ estimate } \theta(F), \text{ calling the estimate } \hat{\theta}_1.$$
From the second sample, say
$$Y_{n+1}, Y_{n+2}, \ldots, Y_{2n}, \text{ estimate } \theta(F) \text{ by } \hat{\theta}_2.$$
$$\vdots$$
From the Nth sample, say
$$Y_{(N-1)n+1}, \ldots, Y_{Nn}, \text{ estimate } \theta(F) \text{ by } \hat{\theta}_N.$$

<u>Step GB-6.</u> The sample $\hat{\theta}_1, \hat{\theta}_2, \ldots, \hat{\theta}_N$ is used to assess the estimate $\hat{\theta}$. [E.g., upper and lower $\alpha/2$ percentiles might be used as a CI.]

Step GB-7. Sometimes a function of $\hat{\theta}_1, \hat{\theta}_2, \ldots, \hat{\theta}_N$ will be used to estimate $\theta(F)$. [In this case $\hat{\theta}$ will not be used as an estimate.]

We have the obvious theorem

Theorem: The Bootstrap is a special case of the Generalized Bootstrap $GB(\theta, F, X_1, \ldots, X_n, F_c, \hat{F}, n, N)$ wherein one sets $\hat{F} = F_n$.

3. AN EXAMPLE OF GB AND COMPARISONS

Consider the problem (given in Karian and Dudewicz (1991)) of estimating θ when F is uniform on $(0, \theta)$ and

$$\{X_1, X_2, \ldots, X_{10}\} = \{0.776, 4.705, 3.333, 5.144, 6.311,$$
$$0.443, 7.337, 4.726, 7.034, 7.119\} \tag{6}$$

is a random sample of size n=10. The MLE of θ is $\max(X_1, X_2, \ldots, X_n)$, so one might estimate θ by

$$\hat{\theta} = \max(X_1, X_2, \ldots, X_n) = 7.337.$$

Suppose the problem were so complex (which, for purposes of illustration, it is not) that an exact CI for θ based on $\hat{\theta}$ were not available. Then The Generalized Bootstrap $GB(\theta, F, X_1, \ldots, X_n, F_c, \hat{F}, n=10, N)$ of Section 2 above might be used.

Example 1: GB with $\hat{F}$ the Empiric p.d.f. The empiric p.d.f. of order 1, introduced in Dudewicz and van der Meulen (1987), is a natural density estimate for the continuous case (and is the natural continuous analog of the empirical d.f.). For the data set (6), choosing bounds of a=0 and b=10 on the range of X, it can be shown that the empiric p.d.f. sets

$$\hat{f}(x) = \begin{cases} 0.164, & 0.000 \le x \le 0.610 \\ 0.069, & 0.610 < x \le 2.054 \\ 0.051, & 2.054 < x \le 4.019 \\ 0.144, & 4.019 < x \le 4.716 \\ 0.455, & 4.716 < x \le 4.935 \\ 0.126, & 4.935 < x \le 5.728 \\ 0.106, & 5.728 < x \le 6.673 \\ 0.248, & 6.673 < x \le 7.076 \\ 0.661, & 7.076 < x \le 7.228 \\ 0.036, & 7.228 < x \le 10.00 \\ 0.000, & \text{otherwise.} \end{cases} \tag{7}$$

Generating N=50 random samples of size n=10 from $\hat{f}$ given in (7), we find

$$\hat{\theta}_1 = 9.724, \quad \hat{\theta}_2 = 9.802, \quad \ldots, \quad \hat{\theta}_{50} = 9.835 \text{ with } s = 1.3284 \tag{8}$$

where s is the sample standard deviation, so an approximate 95% CI for θ is

$$\hat{\theta} \pm 1.96s = (4.733, 9.941). \tag{9}$$

Example 2: GB with $\hat{F}$ a Fitted GLD. The Generalized Lambda Distribution sets

$$F_c^{-1}(y) = \lambda_1 + \frac{y^{\lambda_3} - (1-y)^{\lambda_4}}{\lambda_2}, \tag{10}$$

or

$$f_c(x) = \frac{\lambda_2}{\lambda_3 y^{\lambda_3 - 1} + \lambda_4 (1-y)^{\lambda_4 - 1}}, \tag{11}$$

for appropriate $\lambda_1, \lambda_2, \lambda_3, \lambda_4$ which match the mean, variance, skewness, and kurtosis of F_c to those of the sample $X_1, X_2, \ldots, X_n$. In our case the sample moments are $\overline{X} = 4.6928$, $s = 2.3755$, $\alpha_3 = -0.6671$, $\alpha_4 = 2.1522$ hence (using the tables of Ramberg, Tadikamalla, Dudewicz, and Mykytka (1979)) we take $\lambda_1 = 7.84984$, $\lambda_2 = 0.094296$, $\lambda_3 = 0.4318$, $\lambda_4 = 0.003908$. (While the GLD family can match all $\overline{X}$ and s, it can match (α_3, α_4) only if $\alpha_4 \geq 1.8 + 1.7\alpha_3^2$; when violated, one takes the closest tabled pair to find the λ_i's.) Generating N=50 random samples of size n=10 from F_c^{-1} with the matched λ_i's, we find

$$\hat{\theta}_1 = 7.475, \ \hat{\theta}_2 = 7.625, \ \ldots, \ \hat{\theta}_{50} = 7.688 \text{ with } s = 0.4016 \tag{12}$$

so an approximate 95% CI for θ is

$$\hat{\theta} \pm 1.96s = (6.550, 8.124). \tag{13}$$

Example 3: GB with $\hat{F} = F_n$ (the traditional "Bootstrap"). Sampling from the empirical d.f. for N=50 samples of size n=10 yields

$$\hat{\theta}_1 = 7.119, \ \hat{\theta}_2 = 7.337, \ \ldots, \ \hat{\theta}_{50} = 7.337 \text{ with } s = 0.3196 \tag{14}$$

so an approximate 95% CI for θ is

$$\hat{\theta} \pm 1.96s = (6.711, 7.963). \tag{15}$$

(Note: If one used the CI obtained by trimming 2.5% of the $\hat{\theta}_i$'s at the lower and upper ends, the interval (7.034, 7.337) would result.)

Comparisons. The true value of θ used to generate the data (6) was $\theta=8$. Hence the Empiric p.d.f., and the Fitted GLD, succeed in producing valid CI's (9) and (13) via the Generalized Bootstrap. The Fitted GLD produces a shorter interval (which, however, still includes the true value of θ). The special case of the Generalized Bootstrap GB known as the Bootstrap, yields an interval (15) which fails to cover θ.

While replications of this example could show the true coverage probability for each of the three methods (GB with: Empiric p.d.f., Fitted GLD, Empirical d.f.), our point is made well by the example to this point. Namely, distributions known to be continuous are better handled by a continuous fitted distribution (of which two robust versions have been illustrated), and not with the traditional bootstrap's F_n (the empirical d.f., which is not continuous).

One may wonder if F_n might not be acceptable in light of theory that has been developed to show it gives asymptotically valid results in some cases. In our consulting problems rarely does large n arise, and more often n is very small (due to the expense of the data). Hence asymptotics seem a poor substitute for good small-sample performance. (One might say: If a procedure can't do well with n→∞, it is a poor procedure indeed. The real test is for small n.) As Efron (1982, p. 36) states (in a slightly different context), all "...depends on $\hat{F}$ being a reasonable estimate of F...". The Generalized Bootstrap GB allows this; the Bootstrap with $\hat{F} = F_n$ does not.

4. CONCLUSIONS

We conclude the Generalized Bootstrap introduced in Section 2 is superior in that it fits F with an $\hat{F}$ that incorporates properties known for F (continuity, known range, etc.). We conjecture that asymptotic properties of GB will show it a viable competitor of the traditional bootstrap of Efron (which takes $\hat{F} = F_n$), and that small-sample studies will show GB to be superior in the most important arena in practice.

REFERENCES

Csörgö, S. and Mason, D.M. (1989). Bootstrapping empirical functions. Annals of Statistics, 17, 1447-1471.

Dudewicz, E.J. and Mishra, S.N. (1988). Modern Mathematical Statistics, John Wiley & Sons, Inc., New York.

Dudewicz, E.J. and van der Meulen, E.C. (1987). The empirical entropy, a new approach to nonparametric entropy estimation. In New Perspectives in Theoretical and Applied Statistics (ed. M.L. Puri, J.P. Vilaplana, and W. Wertz), John Wiley & Sons, Inc., New York, pp. 207-227.

Edgington, E.S. (1986). Randomization Tests. In Encyclopedia of Statistical Sciences, Volume 7 (ed. S. Kotz, N.L. Johnson, and C.B. Read), John Wiley & Sons, Inc., New York, pp. 530-538.

Efron, B. (1982). The Jackknife, the Bootstrap and Other Resampling Plans. Vol. 38, CBMS-NSF Regional Conference Series in Applied Mathematics, Society for Industrial and Applied Mathematics, Philadelphia, Pennsylvania.

Gentle, J.E. (1985). Monte Carlo Methods. In Encyclopedia of Statistical Sciences, Volume 5 (ed. S. Kotz, N.L. Johnson, and C.B. Read), John Wiley & Sons, Inc., New York, pp. 612-617.

Hinkley, D. (1983). Jackknife Methods. In Encyclopedia of Statistical Sciences, Volume 4 (ed. S. Kotz, N.L. Johnson, and C.B. Read), John Wiley & Sons, Inc., New York, pp. 280-287.

Karian, Z.A. and Dudewicz, E.J. (1991). Modern Statistical, Systems, GPSS Simulation: The First Course, Computer Science Press/W.H. Freeman and Company, Publishers, New York.

Quenouille, M. (1949). Approximate tests of correlation in time series. Journal of the Royal Statistical Society, 11B, 18-84.

Ramberg, J.S., Tadikamalla, P.R., Dudewicz, E.J., and Mykytka, E.F. (1979). A probability distribution and its uses in fitting data. Technometrics, 21, 201-214.

Simon, J.L. (1969a). The Mathematics Teacher, April 1969.

Simon, J.L. (1969b). Basic Research Methods in Social Science (Third Edition with P. Burstein, 1985).

Swanepoel, J.W.H. (1985). Bootstrap selection procedures based on robust estimators (with discussion). In The Frontiers of Modern Statistical Inference Procedures (ed. E.J. Dudewicz), American Sciences Press, Inc., Columbus, Ohio, pp. 45-64.

Tukey, J. (1958). Bias and confidence in not quite large samples. Abstract, Annals of Mathematical Statistics, 29, 614.

SIMULATION-BASED MULTIPLE COMPARISONS: BACKGROUND, PHILOSOPHY, AND
EXTENSIONS TO THE MULTIVARIATE GENERAL LINEAR MODEL

Don Edwards, Department of Statistics
University of South Carolina
Columbia, SC 29208
USA`

Jack J. Berry
SAS Institute, Inc.
Cary, NC 27511
USA

1. Notation; problem definition

The most widely-used version of the multivariate general linear model (MGLM) supposes that p measurements are made on subject i, denoted here $Y_i' = (Y_{i1}, \ldots, Y_{ip})$, $i=1,\ldots,n$. If Y is the nxp matrix whose ith row is Y_i', it is hypothesized that $Y = X\beta + \varepsilon$, where X is an nxq matrix of known constants, β is a qxp matrix of unknown constants, and ε is an nxp matrix of random variables whose rows ε_i', $i=1,\ldots,n$ are independent, each with mean 0 and positive definite covariance matrix Σ. For exact inference, it is assumed that each ε_i' has a multivariate normal distribution, but the methods described here will be robust to moderate departures from this assumption for moderate to large n. It is also assumed without loss of generality that X is of full rank q.

The goal discussed here is construction of simultaneous $(1-\alpha) \times 100\%$ confidence intervals for k estimable double linear compounds $\theta_1 = c_1'\beta m_1, \theta_2 = c_2'\beta m_2, \ldots,$ $\theta_k = c_k'\beta m_k$, for prespecified constant vectors $c_j \in R^q$ and $m_j \in R^p$, $j=1,\ldots,k$. "Estimability" means that each $c_j = X'a_j$ for some $a_j \in R^n$. Define C as the kxq matrix whose jth row is c_j', and M as the kxp matrix whose jth row is m_j'. Denote the usual unbiased estimators of β, Σ and θ_j by $\hat{\beta}$, S, and $\hat{\theta}_j = c_j'\hat{\beta}m_j$. The degrees of freedom associated with S are denoted ν.

Simulation will be used to accomplish the goal. In §2, the basic results and philosophy of what we call "simulation-based inference" are discussed in the context of the univariate GLM. In §3, four conservative methods for the MGLM are defined; In §4, the relative efficiencies of these are briefly discussed. Unfortunately, space restrictions do not allow an example or details of proofs.

2. Univariate simultaneous estimation via simulation

The univariate GLM is a subcase of the MGLM defined above, for p=1; the m-vectors are now scalars and can be taken =1. To construct simultaneous $(1-\alpha) \times 100\%$ confidence intervals for $\theta_j = c_j'\beta$, $j=1,\ldots,k$, consider the pivotal quantity

$$W = \max_{1<j<k} [\,|c_j'(\hat{\beta}-\beta)|(\hat{\sigma}^2 c_j'(X'X)^{-1}c_j)^{1/2}] = \max_{1<j<k} [\,|\hat{\theta}_j-\theta_j|/S_j], \qquad [1]$$

where $\hat{\sigma}^2$ is the scalar version of S, and S_j is the usual estimated standard error of $\hat{\theta}_j$. If w_{α}, the upper-α quantile of the distribution of W, can be found, then simultaneous intervals for the θ_j are of the form $\hat{\theta}_j \pm w_{\alpha}S_j$, j=1,..,k.

The problem is that w_{α} cannot be found analytically or numerically (at present) for many cases of C and X. Notable exceptions give rise to the Tukey(1953) method for all pairwise comparisons, and the Dunnett(1955,1964) method for comparisons with a control. For most other choices the problem has been dealt with in practice using somewhat wasteful probability inequalities, e.g. the methods of Sidak, Bonferroni, and Scheffé (see e.g. Hochberg and Tamhane(1987)). Instead of these, as a working solution to be used only when w_{α} is unavailable, we substitute a random variable W_{α} for w_{α} such that the resulting intervals have exact coverage probability (1-α) computed over the joint distribution of $\hat{\beta}$, $\hat{\sigma}$, and W_{α}. This W_{α} can be obtained via simulation, according to Lemma 1:

Lemma 1 Let $W_1,..,W_m$ and W be independent random variables from a continuous distribution function G. For specified α, 0 < α < 1, let r=(m+1)(1-α) choosing α and m such that r is an integer. Let W_{α} be the rth smallest of $W_1,..,W_m$. Then P[W > W_{α}] = α.

The proof of Lemma 1 follows easily from the fact that $(W,W_1,..,W_m)$ is exchangeable; $W_1,..,W_m$ are obtained by simulating W. The simple idea behind Lemma 1 has been independently discovered many times, for example by Dwass(1957), Barnard(1963), and Birnbaum(1974). Foutz(1981) was first to suggest it for multiple comparisons, but its application extends well beyond this area. In particular, it is the basis for "Monte Carlo tests" (see e.g. Jockel(1986); Ripley(1987)).

There remains the question of choice of simulation size m, and it is here where we part company with some proponents of Monte Carlo tests. Even recently, recommended simulation sizes have varied, according to author, from under 100 to tens of thousands. We feel very strongly that it should be in the latter order of magnitude. Since W_{α} is an ancillary statistic, use of Lemma 1 (for any simulation size) technically violates the Principle of Ancillarity. The simulation size should be chosen to insure that this violation is of a trivial nature (for analogy, rounding data is a trivial violation of the Likelihood Principle). The "actual" error rate one obtains using Lemma 1 (the conditional error probability, given the simulation) is a random variable. Using Lemma 1 only guarantees that the mean of this random variable is α. The simulation size should be chosen to assure with high probability that this "actual" error rate is very close to the nominal α; Lemma 2 below, from Edwards(1985) but alluded to by several earlier authors, will help to achieve this more stringent goal:

Lemma 2: In the setting of Lemma 1, let $\bar{G}=1-G$. The conditional error rate of Monte Carlo tests / simulation based confidence intervals as a random variable is $\bar{G}(W_\alpha)$, and is distributed as a Beta random variable with parameters $m-r+1$ and r. That is, its density function is

$$\{\Gamma(m+1)/\Gamma(m-r+1)\Gamma(r)\}u^{m-r}(1-u)^{r-1}, \quad 0 < u < 1.$$

$E[\bar{G}(W_\alpha)] = \alpha$, $\text{Var}\{\bar{G}(W_\alpha)\} = \alpha(1-\alpha)/(m+2)$, and for large m, $\bar{G}(W_\alpha)$ is essentially normally distributed.

The proof of Lemma 2 uses the Probability Integral Transform and well-known results on the distribution of order statistics. Figure 1 shows the distribution of the conditional error rate with nominal $\alpha=0.05$ and $m=99$; the vertical bars show the location of the central 99% of the corresponding distribution when $m+1=13,000$. The error rate using $m=99$ will be outside the range $(0.02, 0.09)$ in one out of ten simulations. The difference between $\alpha=0.02$ and $\alpha=0.09$ cannot be considered trivial: as one argument, in the case of a single θ, using Student's t at moderate to large degrees of freedom, one needs twice the sample size to achieve a 98% confidence interval the same (expected) length as a 91% interval.

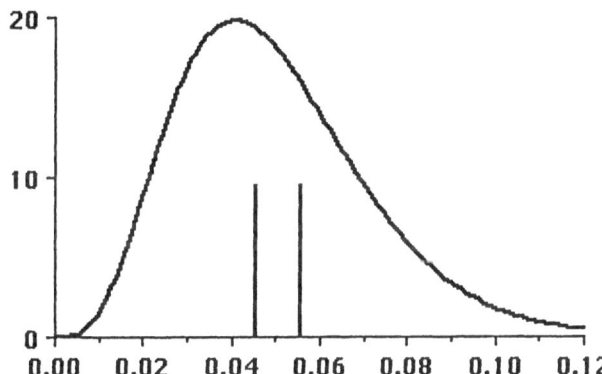

Figure 1: Distribution of the conditional error probability of a Monte Carlo test, m=99, α =.05. The vertical lines at center show the central 99% probability range when m+1=13,000

For a nominal $\alpha=0.05$, we recommend using $m+1=13,000$ (at least). With this choice, one can be 99% confident that the conditional error rate is within 0.005 of $\alpha=0.05$. The CPU time for $m+1=13,000$ on a (slow) IBM 3081 mainframe is about 20 seconds, a small price to pay for sample size savings of 3-19% over the Sidák method, 4-37% over the Bonferroni method, and 27-33% over the Scheffe method, as found by Edwards and Berry(1987) in a comprehensive efficiency study (using $m+1=80,000$). Examples and more details can be found there and in Edwards(1985).

3. Some multiple comparison methods for the Multivariate case

In the MGLM, the natural candidate for a pivotal quantity for simultaneous confidence intervals is

$$W_\Sigma = \max_{1 < j < k} \{ |\hat{\theta}_j - \theta_j| / S_j \},$$

where $S_j = [m_j' S m_j c_j'(X'X)^{-1}c_j]^{1/2}$ again denotes the estimated standard error of $\hat{\theta}_j$. If the upper-α quantile of the distribution of W_Σ could be found, it would give simultaneous intervals

$$\hat{\theta}_j \pm w_{\alpha,\Sigma} S_j, \qquad j=1,..,k. \qquad\qquad [2]$$

Unfortunately, the distribution of W_Σ depends on Σ, as the subscript reminds, so intervals of the form [2] are not available. In this section we sketch four conservative methods, two using the principles of Scheffe's univariate method, and two using variants of the Bonferroni inequality.

3.2 Maximization (Scheffe-type) methods

The first method, called the sim-max method, is "new" and will always give tighter intervals than the well-known Roy and Bose(1953) method for multiple comparisons in an MGLM. For constant c:qx1 and m:px1, let

$$T(c,m) = (c'\hat{\beta}m - c'\beta m) / [m'Smc'(X'X)^{-1}c]^{1/2}$$

Any such $T(c,m)$ has a t-distribution with r degrees of freedom, and $W_\Sigma = \max_j |T(c_j,m_j)|$. Define E_M as the vector space generated by $m_1, m_2, .., m_k$, and let $W_j = \sup\{ |T(c_j,m)| : m \epsilon E_M \}$, $j=1,..,k$. $W_{sm} = \max_j\{W_j\}$ will be the pivotal quantity of the sim-max method. By Lemma 3 below, the distribution of W_{sm} does not depend on unknown parameters. Since each $W_j \geq T(c_j,m_j)$, clearly $\max_j\{W_j\} = W_{sm} \geq W_\Sigma$, so intervals of the form [2] using the upper-α quantile of the distribution of W_{sm} (or the simulation-based version of it) in place of $w_{\alpha,\Sigma}$ will be conservative. Lemma 3 also shows an obvious way to simulate W_{sm}.

Lemma 3 Let r_M = rank (M) and $f_j' = c_j'(X'X)^{-1}X'$. For fixed j, W_j^2 is equal in distribution to

$$\nu f_j' Z' H^{-1} Z f_j / f_j' f_j,$$

where H ~ Wishart(ν,I) of order r_M and Z is an independent $r_M$xn matrix of i.i.d. standard normal variables.

Sketch of proof: the supremum $T^2(c_j,m)$ over E_M can be evaluated by the Cauchy-Schwarz inequality. The resulting expression is a quadratic form in $(\hat{\beta}-\beta)$ and S, which can be shown equal in distribution to the numerator of the above ratio by standard multivariate normal manipulations.

The sim-max method is so named because it maximizes with respect to the m_j vectors, and then applies simulation-based methods to the resulting pivotal. In a few cases of C and X simulation may not be necessary, but finding exact percentiles for the random variable W_{sm} will often be intractable (see for example the attempt by Higazi and Dayton(1984) for balanced ANOVA where the c_j specify comparisons with a control).

An alternative to simulation is to further maximize W_j over the space E_C generated by the c_j's; the corresponding random variable W_{mm} is a monotone transformation of S.N.Roy's greatest root random variable, and using its upper-α percentile as the critical point in [2] is equivalent to the Roy and Bose(1953) method for simultaneous confidence intervals in an MGLM. Clearly the sim-max and Roy-Bose methods will provide increasingly conservative confidence intervals. As it turns out, they are both very data-wasteful.

3.2 Bonferroni-based methods

Two other approaches for the MGLM use the Bonferroni inequality. Typically there will be only a few (say, u) unique m_j-vectors, with the θ_j defined using several c_j for each unique m_j. For example: one-way MANOVA with p=2 components, and we desire all pairwise comparisons of q=6 treatment means for each component. There will be k=30 intervals, but only u=2 unique m_j's. Each unique m_j creates a univariate GLM model $Ym_j = X\beta m_j + \varepsilon m_j$, and simultaneous intervals for the group of θ_j's using this m_j can be obtained by the univariate simulation-based method with $\hat{\sigma}^2 = m_j'Sm_j$. The sim-Bon method uses the univariate simulation-based method to obtain $(1-\alpha/u)\times100\%$ simultaneous intervals for all θ_j in each unique-m_j group; the joint coverage probability of all k intervals will then be at least $1-\alpha$ (more generally, the ith unique-m_j group's intervals (i=1,..,u) could be computed at $(1-\alpha_i)\times100\%$ confidence, where $\Sigma\alpha_i=\alpha$).

The fourth method under discussion is simply to apply the Bonferroni inequality across all k intervals, i.e. use the upper-$(\alpha/2k)$ percent point of Student's-t distribution with ν d.f. as the critical point in [2]. Clearly this full Bonferroni method will be more conservative (but much simpler) than the sim-Bon method.

4. Relative efficiencies of the four methods

An efficiency study comparing the four methods above was performed by Berry(1988), briefly summarized as follows: attention was restricted to the balanced 1-way Multivariate Analysis of Variance (MANOVA) with q treatments. The problem considered was to obtain simultaneous confidence intervals for all

pairwise differences between treatments for each of the p components of the
treatment mean vectors; k=p( ). Nominal confidence level was fixed at $1-\alpha =0.95$.
The structure of Σ was restricted to be autoregressive of order 1, i.e. $\sigma_{ij} = \gamma^{|i-j|}$ for $1\leq i\neq j \leq p$ for varying γ. The study was a $3^2 \times 2^2$ design, varying factors γ
(levels 0.2, 0.5, 0.8), p (levels 2, 3, 6), q (levels 3, 6) and ν (levels 18, 96).

For each of these 36 settings, critical points for the sim–max, Roy–Bose,
sim–Bon, and full Bonferroni methods were obtained. To estimate sample size loss,
the 95% point for W_Σ was also obtained using the known Σ and the methods of §2.
All simulation sizes for critical-point generation were m+1=80,000. For each of
the 36 settings, a stream of 50,000 tests was used to estimate the true coverage
probabilities of the intervals obtained by the five methods; the standard error of
any estimated coverage probability is at most 0.001. All computing used FORTRAN
with IMSL subroutines on a VAX 750 computer. Approximate sample-size waste of each
of the four methods relative to the W_Σ method was estimated as $1-(W_\Sigma$ critical
point)^2/(other critical point)^2.

Of the four methods useful in practice, the sim–Bon method was superior
across all settings, with estimated coverage probabilities ranging from .949 to
.965 and estimated sample size waste relative to the W_Σ critical point ranging
from 0.0% to 8.9%. The much simpler pure Bonferroni method was close behind, with
estimated coverage probabilities ranging from .955 to .969 and estimated sample
size waste ranging from 3.1% to 13.4%. There was a huge gap in performance between
the Bonferroni methods and the maximization methods. The sim–max method showed
estimated coverage probabilities ranging from .974 to .9998 and estimated sample
size waste ranging from 14.9% to 66.4%. The Roy–Bose method performed most poorly
of all, with estimated coverage probabilities ranging from .981 to .9999 and
sample size wastes from 21.47% to 74.25%.

In summary, though we have not been able to find an exact method for
simultaneous intervals in the MGLM, useful information as to the relative
efficiencies of existing methods has been gained, and a new method, the sim–Bon
method, has been found slightly superior to the pure Bonferroni method. Whether
the gain in efficiency is worth the trouble and computer time required to obtain
the sim–Bon critical point is not clear, but it is clear that the Roy–Bose and
sim–max methods are unacceptably conservative for this problem. This is an
important result in that it should discourage multivariate analysts from studying
in great detail the distributional properties of pivotals like W_{sm}.

REFERENCES

Barnard,G.A. Discussion of Professor Bartlett's paper. Jour.Roy.Statist. Soc. B
25, 294 (1963).
Berry,Jack J. Multivariate Simultaneous Estimation. Ph.D. Thesis, The University
of South Carolina, Department of Statistics, Columbia, S.C. 29208 U.S.A.
(1988).

Birnbaum,Z.W. Computers and unconventional test statistics, in Reliability and
    Biometry (F.Proschan and R.J.Serfling, eds.), SIAM, Philadelphia (1974).
Dunnett,C.W. A multiple comparisons procedure for comparing several treatments
    with a control. Jour.Amer.Statist.Assoc. 50, 1096–1121 (1955).
Dunnett,C.W. New tables for multiple comparisons with a control. Biometrics 20,
    482–491 (1964).
Dwass,M. Modified Randomization tests for nonparametric hypotheses. Ann. Math.
    Statist. 28, 181–187 (1957).
Edwards,Don. Exact simulation-based inference: a survey, with additions. Jour.
    Statist.Comp.Sim. 22, 307–326 (1985).
Edwards,Don; Berry,Jack J. The efficiency of simulation-based multiple compari-
    sons. Biometrics 43, 913–928 (1987).
Foutz,R.V. Simultaneous tests for finite families of hypotheses. Comm.Statist.A:
    11, 1839–1853 (1981).
Higazi,S.M.F.; Dayton,C.M. Comparing several experimental groups with a control
    in the multivariate case. Comm.Statist. C: 13(2), 227–241 (1984).
Jockel,K.H. Finite sample properties and asymptotic efficiency of monte carlo
    tests. Ann.Statist. 14, 336–347 (1986).
Ripley,Brian D. Stochastic Simulation. John Wiley and Sons, New York (1987).
Roy,S.N.; Bose,R.C. Simultaneous confidence interval estimation. Ann.Math.
    ,Statist. 24, 513–536 (1953).
Sidak,Z. Rectangular confidence regions for the means of multivariate Normal
    distributions. Jour.Amer.Statist.Assoc. 62, 626–633 (1967).
Tukey,J.W. The problem of multiple comparisons. Unpublished manuscript. (1953).

Applications of Monte Carlo Methods in Spatial and Image Analysis

B.D. Ripley

Department of Statistics, University of Oxford

1 South Parks Road, Oxford OX1 3TG, UK †

Monte Carlo methods have a long history in spatial statistics, and have often been used very effectively to sidestep problems of analytical or computational intractability. On the other hand, bootstrap and other non-parametric methods have made no impact and are rarely considered. The reasons are immediate but often overlooked. The author was once asked to referee a paper on the spatial organization of monkey troops, in which the positions were recorded every ten minutes. After a page or so on the virtues of "distribution–free tests", these were used to test hypotheses about the dominant male being the centre of the troop's movements. Unfortunately for this study, "distribution–free" tests do make one assumption, independence, there violated in both space and time. More generally, resampling methods depend on at least exchangeability.

Most work in spatial statistics is done with a very small number of datasets, often just one, within which *all* observations are expected to be dependent. Thus there is no way exchangeability can be found except perhaps under a null hypothesis. On the other hand, it is frequently easy to simulate independent realizations from a postulated model and so achieve the required exchangeability between real and synthetic datasets. Similar problems arise in time series. Hartigan (1990) presents the novel idea of working with periodogram ordinates which are asymptotically independent for a stationary regularly spaced series. Freedman (1984), Bose (1988) and Basawa *et al.* (1989) all work with the *innovations* of autoregressive models, again introducing an independent sequence. Neither of these techniques have close analogues in spatial statistics. Two reports on the possible use of bootstrap techniques are Lele (1988) and Possolo (1988). Lele considers the local conditional structure of a Markov random field, and so uses dependent but exchangeable 'innovations'. Possolo invokes weak spatial dependence to gain approximate exchangeablity of well-separated samples.

1. Monte Carlo tests

The interest in Monte Carlo tests in spatial statistics stems from a remark by Barnard (1963) in the discussion of Bartlett's (1963) paper on the spectral analysis of point processes. Although Bartlett was concerned with point processes on the line, spectral analysis is direction–less and so the technique can be applied equally to spatial transects (Ripley, 1978). It was this paper which led me to propose Monte Carlo tests as tests of goodness-of-fit of models of spatial point processes in Ripley (1977). As this was a read paper to the *Royal Statistical Society* the idea got a wide airing, and a number of applications

---

† Work done whilst at the Department of Statistics, University of Strathclyde, Glasgow.

followed, starting with Besag and Diggle (1977). The method was also picked up by geographers, (Cliff and Ord, 1973, §2.7) via Hope (1968).

Barnard suggested that a statistic T is selected, small values of which measures goodness-of-fit to the model. Then we can simulate m patterns from the model and compute the statistics $T_1, \ldots, T_m$ from those patterns. If the null hypothesis is true, $(T, T_1, \ldots, T_m)$ are exchangeable samples and therefore the probability that T is the rth largest or larger is $r/(m+1)$. Some early work (e.g. Marriott, 1979) suggested that m should be large to achieve reasonably precise test levels. Jöckel (1986) gave more precise information from which to choose m. However, in almost all applications the test is only being used as a rough guide to fit, and small values of m (such as 99 or even 19) often suffice. [In such cases P-values are desirable, and Besag and Clifford (1991) show how to use other sampling schemes whose size will depend on the P-value.]

There are two immediate difficulties with Barnard's basic idea. We need a simple (precise) null hypothesis to simulate the model. The usual ideas could be used to reduce composite hypotheses to simple ones (pivotal quantities, conditioning on sufficient statistics for nuisance parameters) but these are rarely applicable with spatial models. Some applications *do* generate simple null hypotheses: Cliff and Ord have a uniform randomization distribution under their null hypothesis of no spatial autocorrelation. Occasionally conditioning *can* be used. Suppose we have two spatial point patterns to test their association. By conditioning on both patterns up to a random motion, we can base inference on random motions (translations, rotations or as appropriate) of one pattern relative to the other. One example of this technique is to two patterns of ants' nests by Harkness and Isham (1983), although I had seen earlier examples. Fine asymptotic points about Monte Carlo tests for composite null hypotheses with estimated nuisance parameters are made by Hall and Titterington (1989).

The other difficulty is that there must be a single test statistic T. Many useful goodness-of-fit tests produce a curve to be compared with a theoretical curve (as does the graphical form of the Kolmogorov test). In particular, my $\hat{K}$ statistic for second-moment characteristics of spatial point patterns (Ripley 1977) allows the comparison of fit at each interpoint distance. Such batteries of simultaneous tests are now widely used (Ripley, 1981; Diggle, 1983) and have proven useful despite qualms about their imprecise statistical properties.

The basic theory of Barnard's test presumes we can generate independent (or exchangeable) samples. Many of the simulation models used for spatial processes are iterative in nature, generating a series of dependent samples at a rapid rate (Ripley, 1987, §4.7; Ripley and Kirkland, 1990). By taking sample patterns sufficiently far apart in the sequence we can generate essentially independent samples. When the simulation process is cheap and the computation of the statistic expensive this is obviously sensible, but sometimes the sizes of the costs will be reversed. For that case Besag and Clifford (1989) consider Monte Carlo tests from temporally dependent samples. An iterative method defines a Markov chain on the state space. In general it will be irreversible, but can be run forwards or backwards. In many cases, including our example below, the chain is reversible, and is run backwards by the same algorithm used for running forwards.

As an illustration of their procedure, consider the following simple spatial model for discrete observations $(X_1, \ldots, X_N)$ on a graph of sites:

$$P(\mathbf{X}) \propto \exp \beta \#\{\text{nhbrs pairs with equal values}\} \tag{1}$$

and suppose we wish to test whether a sample $(X_1, \ldots, X_N)$ was generated by (1) for some unknown β. Clearly this is a canonical exponential family and hence the count, b, is sufficient for β and conditional on b all realizations $\mathbf{X}$ are equally likely for all β. Let $S(b)$ denote the restricted sample space $\{(X_1, \ldots, X_n) \mid \#\{\text{equal nhbr pairs}\} = b\}$. Then the conditional test is of uniformity over $S(b)$. Suppose we have an iterative simulation method on $S(b)$. (We could for example select two sites at random, select random values for them, and accept the change if and only if the new realization is in $S(b)$.)

Let T be any measure of uniformity of $S(b)$. The Besag-Clifford *parallel* method is to run the chain *backwards* for n steps, then forwards n steps, repeating the forward steps m times to reach m patterns and hence m values of T. These m values are exchangeable with T from the data under the null hypothesis.

The *serial* method generates $m+1$ samples with the Mth as the data, with $M \sim U(\{1, \ldots, m+1\})$. Once again there is exchangeability, and the samples can be generated from the data by running the chain both forwards or backwards from the data as sample M. This is clearly more economical in our example.

Monte Carlo methods can also be applied to randomization tests. Mead (1974) proposed a randomization test of spatial transects, of one length scale within twice that scale. The spatial (two-dimensional) generalization has a large randomization distribution, and Besag and Diggle (1977, §5) suggested using a randomly chosen subset of the $(16!)/(4!)^5 = 2,627,625$ combinations. Further (non-spatial) examples of such randomization tests are given by Noreen (1989), and others, including spatial problems, by Manly (1991).

2. Parameter estimation

Many models of spatial interaction have computationally intractable maximum likelihood estimates of their parameters. A classic example is (1). The MLE of β is obtained by equating b to its expected value under β. Unfortunately neither the normalizing constant in (1) nor the expected value are known except in the most special cases. A similar problem arises for spatial point processes. The so–called Strauss model (Kelly and Ripley, 1975) is a density for the distribution for n points within a set D, with

$$p(\mathbf{x}_1, \ldots, \mathbf{x}_n) = ab^n\, c^{y(R)} \tag{2}$$

where $y(R)$ denotes the number of pairs of points less than distance R apart. The normalizing constant a is unknown as a function of (b, c). Even if we condition on n, the constant is still unknown.

Again we have a canonical exponential family, so the MLE of (b, c) satisfies

$$\begin{aligned} n &= E_{b,c}N \\ y(R) &= E_{b,c}Y(R) \end{aligned} \tag{3}$$

if N is varied, and only the second equation for c if N is conditioned on n. Further details of these models are in Ripley (1988, Chapter 4; 1990a). A number of alternative estimation strategies are available, notably pseudo-likelihood, but it is desirable to be able to compute MLE's at least for comparative studies.

One obvious solution is to estimate the right hand side of (3) (or, for (1), the expected value of b) by simulation, replacing the integration needed to evaluate the normalizing constant. These ideas were first used by Penttinen (1984) for the conditioned Strauss model, and exploited by Ripley (1988) using sophisticated iterative simulation techniques. A further variant using stochastic approximation is given by Moyeed and Baddeley (1989). Iterative simulation methods are used to simulate the Strauss process, and so estimate the expected values at (3). Subtle statistical issues are involved in finding good estimates with high computational efficiency. We can for example take frequent (dependent) samples of the statistics $(N, Y(R))$ provided we allow for the dependence in calculating the accuracy of our estimates.

Simple lattice models such as (1) have become popular as priors in Bayesian image analysis (Geman and Geman, 1984; Ripley, 1988, Chapter 5), and this has led to interest in estimating the parameters in imperfectly observed realisations from these models, as part of an empirical Bayes procedure. For concreteness, assume that our graph is part of a two-dimensional lattice, and that the prior is model (1) with k types at each site. Suppose that each type corresponds to a grey level, and that this is observed with independent Gaussian noise variance κ. Let $(Z_1, \ldots, Z_N)$ denote the observed data. The problem is then to estimate β given Z. All the available techniques rely heavily on simulation.

One approach is taken by Veijanen (1990). The pseudo-likelihood estimator of the parameters of an MRF maximizes

$$PL = \prod_{\text{sites}} P(X_s = x_s \mid X_t, t \neq s)$$

which for (1) depends only on the empirical distribution of the number of neighbours of each site of the same colour. However, this distribution is not directly observable from Z. Veijanen's SPLEMD procedure works in two steps. First the sufficient statistic b is estimated from Z by a method not involving β. Then the MRF X is simulated conditional on b, and the pseudo-likelihood estimator from X is used to estimate β. As we saw in §1, the simulation can be done by an iterative scheme.

Another approach is that of Younes (1989), which is again based on an EM method plus stochastic approximation from simulations.

3. Simulation-based Bayesian inference

Much recent interest in spatial statistics has been directed towards the analysis of image data, mainly from a Bayesian perspective. We saw in §2 how Monte Carlo methods were being used to estimate parameters in the prior in an empirical Bayes setting. The use of simulation is much more widespread. The size of the distributions considered is so large that exact calculations are intractable. (There may be as many as 1 million pixels in an image, each a greylevel or classified as one of a small number of types.) Efficient simulation techniques provide a very appealing way to cover the important parts of the sample space.

The general problem will have a description S of a true image, and a (pixel-based) description of the observed image Z. We specify a distribution $P(S)$ over true images and an observation distribution

$P(Z \mid S)$. Then Bayesian inference is based on

$$P(S \mid Z) \propto P(Z \mid S)P(S) \tag{4}$$

This is still a very high-dimensional distribution, so must be summarized in some way. One rather popular choice is to show its mode, the MAP (maximum *a posteriori*) estimator. In most cases the maximization problem is still non-trivial, especially since a number of quite different images may have values of $P(S \mid Z)$ very close to the maximum.

If S describes each pixel as one of a small number k of types or greylevels, for example landuse in remote sensing or levels 0-255 in tomography, we have a combinatorial optimization problem. Much of the development of the stochastic optimization technique known as *simulated annealing* has been developed within this field (e.g. Geman and Geman, 1984). The essence of the idea is very simple: rather than maximize (4) we draw samples from

$$P_\lambda(S \mid Z) \propto P(S \mid Z)^\lambda \tag{5}$$

for large positive λ. Then P_λ concentrates on the maximum as $\lambda \to \infty$, and samples from P_λ are likely to turn up the MAP estimator of S, or at least images with near-maximal $P(S \mid Z)$. Unfortunately, iterative methods are almost invariably needed to sample from P_λ, and these converge very much more slowly for highly-peaked distributions such as P_λ. A practical compromise is to increase λ as the iterative process runs. There are now results which show that the process converges to the MAP estimator if and only if λ grows logarithmically fast (or, rather, slowly). Aarts and Korst (1989) give a recent review. However, to maximize the chance of getting close to the MAP in a fixed length of run, λ should be increased about exponentially fast (Catoni, 1990).

Simulation is also useful with other loss functions. MAP estimation corresponds to a global 0-1 loss. When S describes a classification of each pixel, some authors (e.g. Marroquin *et al.*, 1987) prefer to select the mode for each pixel type, corresponding to minimizing the expected number of misclassifications. Thus for each pixel s we choose the mode of $P(S_s \mid Z)$ which involves summation over $\{S_t, t \neq s\}$. This is impracticable analytically. An idea of Grenander (1983) uses simulation. Draw say, 1000 samples from $P(S \mid Z)$ and select the highest-frequency type at each pixel. With independent samples the modal type would be found almost certainly. However, in almost all cases only iterative simulation methods are available to sample from $P(S \mid Z)$, and it is essential to run the simulation for long enough to evenly cover the sample space.

Very recent work (Chow *et al*, 1989; Ripley, 1990b; Ripley and Sutherland, 1990) considers a higher-level description of S as a collection of objects specified by sketch outlines. Here simulation is used to illustrate the range of possible fitted objects by showing a series of samples from $P(S \mid Z)$.

Simulation-based methods have already proved their value in the computationally demanding field of image analysis. Often no other approach to inference is available, nor is one forseeable. The availability of iterative simulation methods remains the key to the success of this approach.

References

Aarts, E. and Korst, J. (1989). *Simulated Annealing and Boltzmann Machines*. Wiley, Chichester.

Barnard, G. (1963). Contribution to the discussion of Prof. Bartlett's paper. *J. Roy. Statist. Soc. B* **25**, 296.

Bartlett, M.S. (1963). The spectral analysis of point processes. *J. Roy. Statist. Soc. B* **25**, 264-296.

Basawa, I.V., Mallik, A.K., McCormick, W.P. and Taylor, R.L. (1989). Bootstrapping explosive autoregressive processes. *Ann. Statist.* **17**, 1474-1486.

Besag, J. and Clifford, P. (1989). Generalized Monte Carlo significance tests. *Biometrika* **76**, 633-642.

Besag, J. and Clifford, P. (1991). Sequential Monte Carlo p-values. *Biometrika* **78**.

Besag, J. and Diggle, P.J. (1977). Simple Monte Carlo tests for spatial pattern. *Appl. Statist.* **26**, 327-333.

Bose, A. (1988). Edgeworth correction by bootstrap in autoregressions. *Ann. Statist.* **16**, 1704-1772.

Catoni, O. (1990) Rough large deviation estiamtes for simulated annealing: application to exponential schedules. Submitted to *Ann. Probab.*.

Chow, Y., Grenander, U. and Keenan, D.M. (1989). *HANDS. A Pattern Theoretic Study of Biological Shapes*. Division of Applied Mathematics, Brown University.

Cliff, A.D. and Ord, J.K. (1973). *Spatial Autocorrelation*. Pion, London.

Diggle, P.J. (1983). *Statistical Analysis of Spatial Point Patterns*. Academic Press, London.

Freedman, D. (1984). On bootstrapping two-stage least squares estimates in stationary linear models. *Ann. Statist.* **12**, 827-842.

Geman, S. and Geman, D. (1984). Stochastic relaxation, Gibbs distributions and the Bayesian restoration of images. *IEEE Trans PAMI* **6**, 721-741.

Grenander, U. (1983). *Tutorial in Pattern Theory*. Division of Applied Mathematics, Brown University.

Hall, P.J. and Titterington, D.M. (1989). The effect of simulation order on level accuracy and power of Monte Carlo tests. *J. Roy. Statist. Soc. B* **51**, 459-467.

Harkness, R.D. and Isham, V. (1983). A bivariate spatial point pattern of ants' nests. *Appl. Statist.* **32**, 293-303.

Hartigan, J.A. (1990). Perturbed periodogram estimates of variance. *Int. Statist. Rev.* **58**, 1-7.

Hope, A.C.A. (1968). A simplified Monte Carlo significance test procedure. *J. Roy. Statist. Soc. B* **30**, 582-598.

Jöckel, K.-H. (1986). Finite sample properties and asymptotic efficiency of Monte Carlo tests. *Ann. Statist.* **14**, 336-347.

Kelly, F.P. and Ripley, B.D. (1975). A note on Strauss's model for clustering. *Biometrika* **63**, 357-360.

Lele, S. (1988). Non-parametric bootstrap for spatial processes. Abstract 206-77 *Bull. Int. Math. Statist.* **17**, 237.

Manly, B.F.J. (1991) *Randomization and Monte Carlo Methods in Biology*. Chapman and Hall, London.

Marriott, F.H.C. (1979). Barnard's Monte Carlo tests: how many simulations? *Appl. Statist.* **28**, 75-77.

Marroquin, J., Mitter, S. and Poggio, T. (1987). Probabilistic solution of ill-posed problems in computational vision. *J. Amer. Statist. Assoc.* **82**, 76-89.

Mead, R. (1974). A test for spatial patterns at several scales using data from a grid of contiguous quadrats. *Biometrics* **30**, 295-307.

Moyeed, R.A. and Baddeley, A.J. (1989). Stochastic approximation of the MLE of a spatial point pattern. Report BS-R8926, CWI, Amsterdam.

Noreen, E.W. (1989). *Computer Intensive Methods for Testing Hypotheses: An Introduction.* Wiley, New York.

Penttinen, A. (1984). Modelling spatial interaction in spatial point patterns: parameter estimation by the maximum likelihood method. *Jyväskylä Studies in Computer Science, Economics and Statistics* **7**.

Possolo, A. (1988). Subsampling and resampling statistics of spatial processes. Abstract 206-38 *Bull. Int. Math. Statist.* **17**, 227.

Ripley, B.D. (1977). Modelling spatial patterns. *J. Roy. Statist. Soc. B* **39**, 172-212.

Ripley, B.D. (1978). Spectral analysis and the analysis of pattern in plant communities. *J. Ecology* **66**, 965-981.

Ripley, B.D. (1981). *Spatial Statistics.* Wiley, New York.

Ripley, B.D. (1987). *Stochastic Simulation.* Wiley, New York.

Ripley, B.D. (1988). *Statistical Inference for Spatial Processes.* Cambridge University Press, Cambridge.

Ripley, B.D. (1990a). Gibbsian interaction models. In *Spatial Statistics: Past Present and Future* ed D.A. Griffith, IMAGE, Ann Arbor, 3-25.

Ripley, B.D. (1990b). Recognizing organisms from their shapes – a case study in image analysis. *Proc. XVth International Biometrics Conference, Budapest*, Invited Papers pp. 259-263.

Ripley, B.D. and Kirkland, M.D. (1990). Iterative simulation methods. *J. Comp. Appl. Math.* **31**, 165-172.

Ripley, B.D. and Sutherland, A.I. (1990). Finding spiral structures in images of galaxies. *Phil. Trans. Roy. Soc. A.* **332**, 477-485.

Veijanen, A. (1990). An estimator for imperfectly observed Markov random fields. Research report 764, Dept. of Statistics, University of Helsinki.

Younes, L. (1989). Parametric inference for imperfectly observed Gibbsian fields. *Probab. Theory Related Fields* **82**, 625-645.

Bootstrap Bands for Confidence and Prediction Regions (3)

CONFIDENCE BANDS FOR PROBABILITY DISTRIBUTIONS ON VAPNIK-CHERVONENKIS CLASSES OF SETS IN ARBITRARY SAMPLE SPACES USING THE BOOTSTRAP

Peter Gaenssler

Mathematical Institute, University of Munich
Theresienstraße 39, D-8000 Munich 2

ABSTRACT. The construction of confidence bands for an unknown distribution function (df) on the real line $\mathbb{R}$ using Efron's (1979) bootstrap procedure is well known through the work of Bickel and Freedman (1981). It is based on a central limit theorem for the bootstrapped empirical process on $\mathbb{R}$, which has been subsequently generalized together with parallel results for empirical processes based on observations in $\mathbb{R}^d$, $d > 1$, and in arbitrary sample spaces, where the corresponding processes are then indexed by certain classes of sets or functions. The present state of this subject will be shortly reviewed. Special attention is given to empirical processes indexed by Vapnik-Chervonenkis classes $\mathscr{C}$ of sets in arbitrary sample spaces together with the construction of confidence bands for probability distributions on $\mathscr{C}$ as in the classical case.

AMS 1980 subject classifications. Primary 62G15, Secondary 60F17.
Key words and phrases. Empirical processes, bootstrapped versions, functional central limit theorems, Vapnik-Chervonenkis classes, confidence bands, critical values determined by bootstrapping.

INTRODUCTION. Let $(\xi_i)_{i\in\mathbb{N}}$ be a sequence of i.i.d. random elements (re's) defined on some basic p-space $(\Omega,\mathscr{A},\mathbb{P})$ with values in an arbitrary sample space $(X,\mathfrak{X})$ (i.e. $\xi_i: \Omega \to X$ is assumed to be $\mathscr{A}$, $\mathfrak{X}$-measurable) and with distribution μ on $\mathfrak{X}$ ($\mu(B) = \mathbb{P}(\xi_i\in B)$, $B\subset \mathfrak{X}$). Let μ_n be the empirical measure based on $\xi_1,...,\xi_n$, i.e. $\mu_n(B) \equiv n^{-1}\sum_{i=1}^{n} 1_B(\xi_i)$, $B\in\mathfrak{X}$, where 1_B denotes the indicator function of B. Given a subclass $\mathscr{C}$ of $\mathfrak{X}$, the empirical $\mathscr{C}$-process $\beta_n = (\beta_n(C))_{C\in\mathscr{C}}$ (being a stochastic process indexed by $\mathscr{C}$) is defined by $\beta_n(C) \equiv n^{1/2}(\mu_n(C)-\mu(C))$, $C\in\mathscr{C}$.

Next, given a realization $\underline{x} \equiv (x_1,x_2,...)$ of $\underline{\xi} \equiv (\xi_1,\xi_2,...)$ and sample sizes n and m = m(n), respectively, following Efron's nonparametric bootstrap procedure, let $\mu_n^x(C) \equiv n^{-1}\sum_{i=1}^{n} 1_C(x_i)$ and $\mu_{m,n}(C) \equiv m^{-1}\sum_{i=1}^{m} 1_C(\xi_i^*)$, the $\xi_1^*,...,\xi_m^*$ being i.i.d. re's in $(X,\mathfrak{X})$ with distribution μ_n^x on $\mathfrak{X}$ and defined on the same p-space $(\Omega,\mathscr{A},\mathbb{P})$ (properly enlarged). Then $\beta_{m,n}^x = (\beta_{m,n}^x(C))_{C\in\mathscr{C}}$ with $\beta_{m,n}^x(C) \equiv m^{1/2}(\mu_{m,n}(C)-\mu_n^x(C))$, $C\in\mathscr{C}$, is the so-called bootstrapped empirical $\mathscr{C}$-process.

The basic paper by Dudley (1978) was the starting point for the development of a general limit theory for empirical $\mathscr{C}$-processes showing as a main result that (under some measurability assumptions) in case of Vapnik-Chervonenkis classes (VCC) $\mathscr{C}$ a functional central limit theorem (FCLT) holds, i.e.

(1) $\beta_n \xrightarrow{\mathscr{L}} \mathbb{G}_\mu$ as n→∞, where $\mathbb{G}_\mu = (G_\mu(C))_{C \in \mathscr{C}}$ is a mean-zero Gaussian process with cov $(G_\mu(C_1), G_\mu(C_2)) = \mu(C_1 \cap C_2) - \mu(C_1)\mu(C_2)$.

Here $\xrightarrow{\mathscr{L}}$ denotes convergence in law (weak convergence) in a suitable sense, the most general and appropriate one, proposed by Hoffmann-Jørgensen (1984), can be found e.g. in Dudley (1985).

Recall that $\mathscr{C} \subset \mathfrak{X}$ is a VCC iff there exists an s∈ℕ such that $m^{\mathscr{C}}(s) < 2^s$, where $m^{\mathscr{C}}(s) \equiv \max\{\Delta^{\mathscr{C}}(F): F \subset X$ and $|F| = s\}$ with $\Delta^{\mathscr{C}}(F) \equiv |\{F \cap C: C \in \mathscr{C}\}|$. See Dudley (1984), Pollard (1984) and Stengle and Yukich (1989) for examples of VCC's (which in many cases are also measure determining classes); to mention only a few here: In $\mathbb{R}^d$, d ≥1, the classes consisting of all lower-left orthants, all rectangles, or all closed Euclidean balls, respectively, are VCC's. Vapnik and Chervonenkis (1971) introduced the above quantities $\Delta^{\mathscr{C}}$ and $m^{\mathscr{C}}$ measuring the 'size' of classes $\mathscr{C}$ of sets in arbitrary sample spaces X. Now, concerning the bootstrapped version $\beta_{m,n}^X$ of β_n it was shown in Gaenssler (1986) (cf. also Gaenssler and Stute (1987)) that for almost all (a.a.) x

(2) $\beta_{m,n}^X \xrightarrow{\mathscr{L}} \mathbb{G}_\mu$ if m=m(n) → ∞ as n→∞ .

Specializing (1) and (2) to X = ℝ and $\mathscr{C} \equiv \{(-\infty, t]: t \in \mathbb{R}\}$ (being the simplest VCC in ℝ) one meets the classical case with (1) being Donsker's FCLT for empirical processes and with (2) being its bootstrapped analogon as proved first in Bickel and Freedman (1981); cf. also Shorack (1982) for a short proof. We also refer to Beran, Le Cam and Millar (1987), where in Section 5B bootstrapping is sketched for general VCC's.

A SHORT REVIEW ON RESULTS FOR EMPIRICAL AND BOOTSTRAPPED EMPIRICAL PROCESSES. Limiting theory for empirical processes based on observations in arbitrary sample spaces X and indexed by a class $\mathscr{F}$ of measurable functions more general than $\{1_C: C \in \mathscr{C}\}$ has grown and developed enormously in the last years, so our short review here must be necessarily incomplete. The papers and books by Dudley (1984, 1985), Giné and Zinn (1984) and Pollard (1984, 1989) are among the major contributions in this direction. Concerning results on bootstrapping empirical processes indexed by $\mathscr{F}$ (i.e. instead of $\beta_n(C)$ one considers $\beta_n(f) \equiv n^{1/2}(n^{-1} \sum_{i=1}^{n} f(\xi_i) - \int f d\mu)$, $f \in \mathscr{F}$) a certain final result is due to Giné and Zinn (1990). They proved (under a certain measurability assumption) that the following two statements (3) and (4) are equivalent:

(3) (a) The empirical process indexed by $\mathcal{F}$ satisfies the FCLT, and

   (b) the envelope function F of $\mathcal{F}$ (i.e. $F \equiv \sup\{|f|: f\in\mathcal{F}\}$) is square integrable w.r.t. μ

(4) For a. a. sample points $\underline{x}$ the bootstrapped empirical process indexed by $\mathcal{F}$ and with sample size $m(n) \equiv n$ satisfies the FCLT.

In Sheehy and Wellner (1988 and 1990) consequences of P-uniform FCLT's (where the empirical processes converge weakly to $\mathbb{G}_\mu$ uniformly for all $\mu\in\mathbf{P}$, $\mathbf{P}$ being a subcollection of all p-measures on $\mathfrak{X}$) are examined for various versions of the bootstrapped empirical process indexed by functions on general sample spaces X. Pursuing further the line of Bickel and Freedman (1981), Csörgö and Mason (1989) studied the validity of the bootstrap for general empirical functions on $X = \mathbb{R}$ containing as special case the empirical df.

CONFIDENCE BANDS FOR PROBABILITY DISTRIBUTIONS VIA BOOTSTRAP. Beran and Millar (1986) obtained asymptotically valid confidence sets for an unknown probability distribution defined on the VCC of all halfspaces in $\mathbb{R}^d$, the critical values being determined by bootstrapping, treating also the nontrivial computational aspects. Concerning empirical processes indexed by VCC's $\mathcal{C}$ in an arbitrary sample space $(X,\mathfrak{X})$ it follows from (2) that for a.a. $\underline{x}$

$$(5) \quad \sup_{C\in\mathcal{C}} |\beta^{\underline{x}}_{m,n}(C)| \xrightarrow{\mathcal{L}} \sup_{C\in\mathcal{C}} |G_\mu(C)|\,,$$

whence for a.a. $\underline{x}$

$$(6) \quad H^{\underline{x}}_m(t) = \mathbb{P}\Big(\sup_{C\in\mathcal{C}} |\beta^{\underline{x}}_{m,n}(C)| \le t\Big) \longrightarrow H_\mu(t) \quad \text{if} \quad m(n)\to\infty \quad \text{as} \quad n\to\infty$$

   for all $t\in\text{cont}\, H_\mu$, where $\text{cont}\, H_\mu$ denotes the set of all continuity points of
   $H_\mu(\cdot) \equiv \mathbb{P}\Big(\sup_{C\in\mathcal{C}} |G_\mu(C))| \le \cdot\Big)$.

Now, given $0 < \alpha < 1$, choose a critical value $c = c(\alpha,\mu)$ such that $c\in\text{cont}\, H_\mu$ and such that $H_\mu(c) = 1-\alpha$, and assume in addition that

$$c = \inf\{t: H_\mu(t) = 1-\alpha\} = \sup\{t: H_\mu(t) = 1-\alpha\}$$

(cf. Beran and Millar (1986), Section 4.2 for the general frame where this assumption is fulfilled due to the fact that H_μ is strictly increasing (and also continuous in many situations)).

Then, defining $c_m = c_m(\underline{x}) \equiv \inf\{t: H^{\underline{x}}_m(t) > 1-\alpha\}$, it follows from (6) (cf. Witting and Nölle (1970), 2.11,p. 53) that for a.a. $\underline{x}$ $c_m(\underline{x}) \to c$ if $m(n) \to \infty$ as $n \to \infty$, which in turn implies by (1) that

$$\sup_{C\in\mathcal{C}} |\beta_n(C)| + c - c_{m(n)}(\underline{\xi}) \xrightarrow{\mathcal{L}} \sup_{C\in\mathcal{C}} |G_\mu(C)| \quad \text{as } n\to\infty\,.$$

Since $c \in \mathrm{cont}\, H_\mu$, we arrive at

$$\mathbb{P}\left(\sup_{C \in \mathscr{C}} |\beta_n(C)| \leq c_{m(n)}(\underline{\xi}) \right) \longrightarrow H_\mu(c) = 1-\alpha \ ,$$

whence a confidence band for $\mu = (\mu(C))_{C \in \mathscr{C}}$ of asymptotic level $1-\alpha$ is given by

$$\underline{x} \longmapsto \{ \mu_n^{\underline{x}}(C) \pm n^{-1/2}\, c_{m(n)}(\underline{x}) \ , \ C \in \mathscr{C} \} \ ,$$

where $c_{m(n)}(\underline{x})$ can be computed (in principle) by Monte Carlo simulation methods.

REFERENCES

Beran, R., Le Cam, L. and Millar, P.W. (1987). Convergence of stochastic empirical measures. J. Multivariate Anal. 23 159-168.

Beran, R. and Millar, P.W. (1986). Confidence sets for a multivariate distribution. Ann. Statist. 14 431-443.

Bickel, P.J. and Freedman, D.A. (1981). Some asymptotic theory for the bootstrap. Ann. Statist. 9 1196-1217.

Csörgö, S. and Mason, D.M. (1989). Bootstrapping empirical functions. Ann. Statist. 17 1447-1471.

Dudley, R.M. (1978). Central limit theorems for empirical measures. Ann. Probability 6 899-929. (Correction (1979), ibid. 7 909-911.)

Dudley, R.M. (1984). A course on empirical processes. Lecture Notes in Math. 1097 2-142, Springer-Verlag, New York.

Dudley, R.M. (1985). An extended Wichura theorem, definitions of Donsker class, and weighted empirical distributions. Lecture Notes in Math. 1153 141-178, Springer-Verlag, New York.

Efron, B. (1979). Bootstrap methods: another look at the jackknife. Ann. Statist. 7 1-26.

Gaenssler, P. (1986). Bootstrapping empirical measures indexed by Vapnik-Chervonenkis classes of sets. Proc. IV Vilnius Conference (1985) Prob. Theory and Math. Stat., Vol 1 467-481, VNU Science Press, Utrecht.

Gaenssler, P. and Stute, W. (1987). Seminar on Empirical Processes. DMV Seminar, Band 9, Birkhäuser Verlag.

Giné, E. and Zinn, J. (1984). Some limit theorems for empirical processes. Ann. Probability 12 929-989.

Giné, E. and Zinn, J. (1990). Bootstrapping general empirical measures. Ann. Probability 18 851-869.

Pollard, D. (1984). Convergence of Stochastic Processes. Springer-Verlag, New York.

Pollard, D. (1989). Empirical Processes· Theory and Applications. Iowa Lecture Notes.

Sheehy, A. and Wellner, J.A. (1988). Uniformity in P of some limit theorems for empirical measures and processes. Technical Report, vol. 134. Department of Statistics, University of Washington, Seattle.

Sheehy, A. and Wellner, J.A. (1990). Uniform Donsker classes of functions. Preprint.

Shorack, G.R. (1982). Bootstrapping robust regression. Comm. Statist. A-Theory Methods 11 961-972.

Stengle, G. and Yukich, J.E. (1989). Some new Vapnik-Chervonenkis classes. Ann. Satist. 17 1441-1446.

Vapnik, V.N. and Chervonenkis, A.Ya. (1971). On uniform convergence of relative frequencies to their probabilities. Theory Probab. Appl. 16 264-280.

Witting, H. and Nölle, G. (1970). Angewandte Mathematische Statistik. B.G. Teubner, Stuttgart.

BOOTSTRAP CONFIDENCE BANDS

Wolfgang HÄRDLE
C.O.R.E.
Voie du Roman Pays 34, B-1348 Louvain-la-Neuve

Michael Nussbaum
Karl-Weierstrass-Institut für Mathematik
Mohrenstr. 39
D-1086 Berlin

Abstract

Bootstrap confidence bands are constructed for nonparametric regression. Resampling is based on a suitably estimated residual distribution. The procedure is called the *Wild Bootstrap*. The method is to construct first a fine grid of error bars with simultaneous coverage probability. Second the end-points of these error bars are joined via polygon pieces or parabolae using assumptions on the local curvature of the regression curve.

1. Motivation

Nonparametric regression smoothing is a flexible method for estimation of mean curves. Since this technique makes no structural assumptions on the underlying curve, it is very important to have a device for understanding when observed features are significant. An often asked question in this context is whether or not an observed peak or valley is actually a feature of the underlying regression function or is only an artifact of the observational noise. For such issues confidence bands should be used.

This paper proposes and analyzes a method of obtaining confidence bands based on simultaneous error bars at a grid of points. The method is simple to implement and relies on local smoothness of the regression curve. The construction is based on a residual resampling technique which models the conditional error distribution and also takes the bias properly into account.

For an understanding of these ideas, consider Figure 1. Figure 1a shows a scatter plot of the expenditure for potatoes as a function of income for the year 1973, from the Family Expenditure Survey (1968-1983). Figure 1b shows a nonparametric regression estimate which was obtained by smoothing the point cloud, using the kernel algorithm described in Section 2. As a means of understanding the variability in the kernel smooth, Figure 1b also shows error bars, constructed by the Wild Bootstrap method proposed in Härdle and Marron (1990). These bars are estimated simultaneous 80 % confidence intervals. Note that the error bars are longer on the right hand side, which reflects the fact that there are fewer observations there, and hence more uncertainty in the curve estimate. The error bars are also asymmetric in particular at points with high curvature which reflects the correct centering of the bars by a bias term. We propose a method of joining these error bars in order to obtain a bootstrap confidence band.

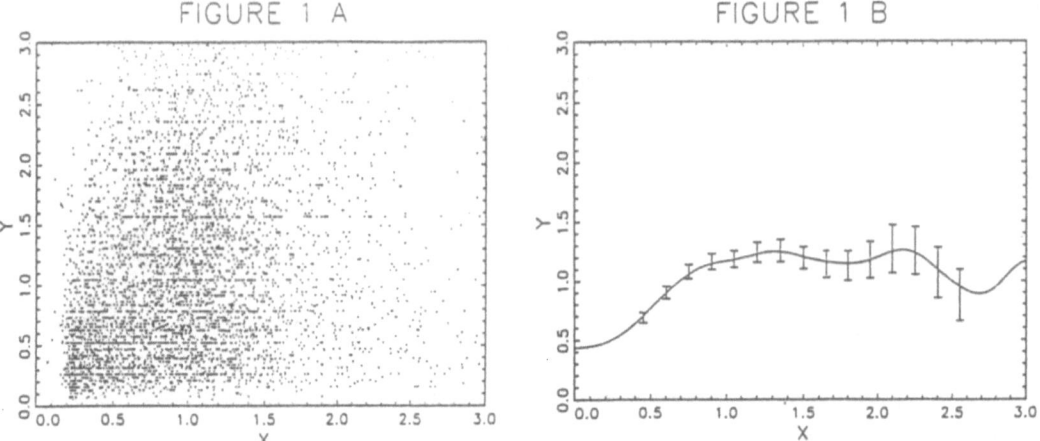

FIGURE 1 A   FIGURE 1 B

Figure 1 a,b. Expenditure for potato (Y) vs. income (X) (a) Scatter Plot (b) Regression kernel smooth (quartic kernel with band with $h=0.3$) and errors bars.

Clearly there is a need for confidence bands in all applications of nonparametric regression. Hall and Titterington (1986) constructed a confidence band for calibration of radio carbon dating assuming Normal errors. Knafl, Sacks and Ylvisaker (1984) derived uniform variability bands under the same error structure.

Our approach is based on resampling from estimated residuals. This form of bootstrapping preserves the error structure in the data and guarantees that the bootstrap observations have errors with mean zero. There are two main advantages to this approach. First it correctly accounts for the bias and hence does not require additional estimation of bias or the use of a sub-optimal (under smoothed) curve estimator. Second, no assumption of homoscedasticity is required, the method automatically adapts to different residual variances at different locations.

In Section 2 we give a technical introduction into simultaneous error bars constructed via the Wild Bootstrap. In Section 3 we consider bootstrap confidence bands. Proofs are given in the forthcoming paper by Härdle and Marron (1990).

2. Simultaneous error bars via the Wild Bootstrap

Stochastic design nonparametric regression is based on iid. observations $\{(X_i, Y_i)\}_{i=1}^n \in \mathbb{R}^{d+1}$ and the goal is to estimate $m(x) = E(Y|X = x) : \mathbb{R}^d \to \mathbb{R}$. The form of the kernel regression estimator we consider is

$$\hat{m}_h(x) = n^{-1} \sum_{i=1}^{n} K_h(x - X_i) Y_i / \hat{f}_h(x) \qquad (2.1)$$

where

$$\hat{f}_h(x) = n^{-1} \sum_{i=1}^{n} K_h(x - X_i) \qquad (2.2)$$

and where $K_h(u) = h^{-d} K(u/h)$ is a kernel weight function with bandwidth h. All results of this paper are stated in terms of this estimator, although the essential ideas clearly extend to other types of kernel estimators such as those of Gasser and Müller (1984) and also to other regression estimators, such as spline methods, as discussed in Eubank (1988).

One approach to the problem of finding simultaneous error bars would be to work with limiting normal distributions of the estimator at the grid points. However the joint distribution of the estimator at close gridpoints has substantial positive correlation, which makes the derivation of joint normal theory confidence intervals nontrivial. In fact, they essentially should be done by simulation methods. Since simulation methods are needed anyway, it seems more economical to use direct resampling.

While bootstrap methods are well known tools for assessing variability, more care must be taken to properly account for the type of bias encountered in nonparametric curve estimation. In particular, the naive bootstrap approach, of resampling from the pairs $\{(X_i, Y_i) : i = 1, ..., n\}$ is inappropriate because the bootstrap bias will be 0, see Härdle and Mammen (CORE DP 9049). Our approach to this problem is to first use the estimated residual

$$\hat{\varepsilon}_i = Y_i - \hat{m}_h(X_i). \qquad (2.3)$$

To better retain the conditional distributional characteristics of the estimate, we do not re-sample from the entire set of residuals, as in Härdle and Bowman (1988). We use the idea of *wild bootstrapping*, as used in Härdle and Marron (1990) where each bootstrap residual is drawn from the two point distribution which has mean zero, variance equal to the square of the residual, and third moment equal to the cube of the residual. In particular define a new random variable ε_i^* having a two point distribution $\hat{G}_i$, where $\hat{G}_i = \gamma \delta_a + (1 - \gamma) \delta_b$ is defined through the three parameters a, b, γ, and where δ_a, δ_b denote point measures at a, b respectively. Some algebra reveals that the parameters a, b, γ at each location X_i are given by $a = \hat{\varepsilon}_i(1 - \sqrt{5})/2$, $b = \hat{\varepsilon}_i(1 + \sqrt{5})/2$ and $\gamma = (5 + \sqrt{5})/10$. These parameters ensure that $E\varepsilon^* = 0, E\varepsilon^{*2} = \hat{\varepsilon}_i^2$ and $E\varepsilon^{*3} = \hat{\varepsilon}_i^3$. The above formulation of the wild bootstrap, based on a two point distribution, is only one possible approach. Other distributions such as mixtures of normals could be considered as well.

After resampling, bootstrap data

$$Y_i^* = \hat{m}_g(X_i) + \varepsilon_i^* \qquad (2.4)$$

are defined, where $\hat{m}_g(x)$ is a kernel estimator with bandwidth g taken to be larger than h (a heuristic explanation of why it is essential to oversmooth g is given below). Then the kernel smoother (2.1) is applied to the bootstrapped data $\{(X_i, Y_i^*)\}_{i=1}^n$ using bandwidth h. Let $\hat{m}_h^*(x)$ denote this kernel smooth. A number of replications of $\hat{m}_h^*(x)$ can be used as the basis for simultaneous error bars because the distribution of $\hat{m}_h(x) - m(x)$ is approximated by the distribution of $\hat{m}_h^*(x) - \hat{m}_g(x)$, as Theorem 1 shows.

For an intuitive understanding of why the bandwidth g used in the construction of the bootstrap residuals should be oversmoothed, consider the asymptotic bias in the case of uniform $f(x)$:

$$E^{Y|X}(\hat{m}_h(x) - m(x)) \approx h^2 (\int u^2 K/2) m''(x),$$

$$E^*(\hat{m}_h^*(x) - \hat{m}_g(x)) \approx h^2 (\int u^2 K/2) \hat{m}_g''(x).$$

Hence for these two distributions to have the same bias we need $\hat{m}_g''(x) \to m''(x)$. This requires choosing g tending to zero at a rate slower than the *optimal bandwidth* h for estimating $m(x)$!

The simultaneous error bars are found as follows. First partition the set of locations where error bars are to be computed into M groups. Suppose the groups are indexed by $j = 1, \cdots, M$ and the locations within each group are denoted by $x_{j,k}, k = 1, \cdots, N_j$. More precisely, the set of grid points $x_{j,k}, k = 1, \cdots, N_j$ has the same asymptotic relative location c_k (not depending on n) to some reference point $x_{j,0}$ in each group j. Therefore define

$$x_{j,k} = c_k h + x_{j,0}, k = 1, \cdots, N_j. \tag{2.5}$$

In the multidimensional case, the simplest formulation is to have each group lying in a hypercube with length $2h$. Now within each group j we use the bootstrap replications to approximate the joint distribution of

$$\hat{m}_h(\underline{x}) - m(\underline{x}) = \{\hat{m}_h(x_{j,k}) - m(x_{j,k}) : k = 1, \cdots, N_j\}.$$

Next we state a theorem which shows that the bootstrap works for the set of locations within each group. For notational convenience we suppress the dependence on j. Technical assumptions are

(1) $m(x), f(x)$ and $\sigma^2(x) = Var(Y|X = x)$ are twice continuously differentiable.

(2) The kernel function K is symmetric and nonnegative, $c_K = \int K^2 < \infty$ and $d_K = \int u^2 K(u) du < \infty$.

(3) $\sup_x E(\varepsilon^3 | X = x) < \infty$.

(4) $f(x_0) \geq \eta > 0$.

Under assumptions (1) and (2) a reasonable choice of h will be in the set

$$H_n = [\underline{c} n^{-1/(4+d)}, \bar{c} n^{-1/(4+d)}], \quad 0 < \underline{c} < \bar{c} < \infty.$$

For this choice of bandwidth the kernel smoother $\hat{m}_h(x)$ is asymptotically optimal, see Section 5.1 of Härdle (1990). The exact specification of the rate of convergence of g is less important for the validity of the following theorem, although it must tend to zero at a rate slower than h. Hence it is assumed that g is chosen from the set

$$G_n = [n^{-1/(4+d)+\delta}, n^{-\delta}], \delta > 0.$$

A fine tuning of the choice of bandwidth g is presented in Theorem 3 of Härdle and Marron (1990).

Theorem 1. *Given the assumptions above, we have along almost all sample sequences and for all $z \in \mathbb{R}^N$*

$$sup_{h \in H_n} sup_{g \in G_n} |P^{Y|X}\{\sqrt{nh^d}[\hat{m}_h(x) - m(x)] < z\}$$
$$- P^*\{\sqrt{nh^d}[\hat{m}_h^*(x) - \hat{m}_g(x)] < z\}| \to 0.$$

The reason that uniform convergence (in h and g) in the above result is important, is that it ensures that the result still holds when h or g are replaced by random data driven bandwidths. For each group j this joint distribution is used to obtain simultaneous $1 - \alpha/M$ error bars that are simultaneous over $k = 1, \cdots, N_j$ as follows. Let $\beta > 0$ denote a generic size for individual confidence intervals. Our goal is to choose β so that the resulting simultaneous size is $1 - \alpha/M$. For each $x_{j,k}, k = 1, \cdots, N_j$ define the interval $I_{j,k}(\beta)$ to have endpoints which are the $\beta/2$ and the $1 - \beta/2$ quantiles of the $(\hat{m}_h^*(x_{j,k}) - \hat{m}_g(x_{j,k}))$ distribution. Then define α_β to be the empirical *simultaneous* size of the β confidence intervals, i.e. the proportion of curves which lie outside at least one of the intervals in the group j. Next find the value of β , denoted by β_j, which makes $\alpha_{\beta_j} = \alpha/M$. The resulting β_j intervals within each group j will then have confidence coefficient $1 - \alpha/M$. Hence by the Bonferroni bound the entire collection of intervals $I_{j,k}(\beta_j), k = 1, \cdots, N_j, j = 1, \cdots, M$ will simultaneously contain at least $1 - \alpha$ of the distribution of $\hat{m}_h^*(x_{j,k})$ about $\hat{m}_g(x_{j,k})$. Thus the intervals $I_{j,k}(\beta_j) - \hat{m}_g(x_{j,k}) + \hat{m}_h(x_{j,k})$ will be simultaneous confidence intervals with confidence coefficient at least $1 - \alpha$. The result of this process is summarized as

Theorem 2. *Define M groups of locations $x_{j,k}, k = 1, \cdots, N_j, j = 1, \cdots, M$ where simultaneous error bars are to be established. Compute uniform confidence intervals for each group. Correct the significance level across groups by the Bonferroni method. Then the bootstrap error bars establish asymptotic simultaneous confidence intervals, i.e.*

$$lim_{n \to \infty} P\{m(x_{j,k}) \in I_{j,k}(\beta_j) - \hat{m}_g(x_{j,k}) + \hat{m}_h(x_{j,k}), \tag{2.6}$$

$$k = 1, \cdots, N_j, j = 1, \cdots, M\} \geq 1 - \alpha$$

As a practical method for finding β_j for each group j we suggest the following "halving" approach (also called a bisection search). In particular, first try $\beta = \alpha/2M$, and calculate α_β. If the result is more than α/M, then try $\beta = \alpha/4M$, otherwise next try $\beta = 3\alpha/4M$. Continue this halving approach until neighboring (since only finitely many bootstrap replications are made, there is only a finite grid of possible β's available) values β_* and β^* are found so that $\alpha_{\beta_*} < \alpha/M < \alpha_{\beta^*}$. Finally take a weighted average of the β_* and the β^* intervals where the weights are $(\alpha_{\beta^*} - \alpha/M)/(\alpha_{\beta^*} - \alpha_{\beta_*})$ and $(\alpha/M - \alpha_{\beta_*})/(\alpha_{\beta^*} - \alpha_{\beta_*})$ respectively.

All of the results in this paper have been stated in terms of the so-called stochastic design model where the regressors X are thought of as realizations of random variables. Since these results are all conditional on $X_1, \cdots, X_n$ our ideas carry over immediately to the case where the $X's$ are fixed and chosen by the experimenter.

In the case of binary regression (dose-response curves, Cox (1970, p.8)), where the response variable Y takes on only the values 0 or 1, there are more natural ways of obtaining bootstrap confidence intervals than those described here. A direct application of our method would give bootstrapped data Y^* which take on values different from 0 and 1. A seemingly more natural approach would be to bootstrap from a Bernoulli distribution with parameter $\hat{m}_g(X_i)$. It is interesting to know how fast the convergence in Theorem 1 takes place. This has been analysed via Berry-Esseen bounds as in Cao-Abad (1989). More precisely Cao-Abad shows that the convergence in Theorem 1 is of order $n^{-2/5}$. Härdle, Huet and Jolivet (1990) consider Edgeworth expansions of the studentized statistic $\sqrt{nh}(\hat{m}_h(x) - m(x))/\widehat{var}x$ and as in Hall (1990) find slightly better rates.

3. Bootstrap Confidence Bands

Once simultaneous error bars have been constructed on a grid of points the extension to uniform confidence bands $[\underline{c}(x), \overline{c}(x)]$ such that

$$P\{\underline{c}(x) \leq m(x) \leq \overline{c}(x) \text{ for all } x\} \approx 1 - \alpha$$

can be done in several ways. One approach is based on bounds on the first derivative $m'(x)$. Let us consider for simplicity just two points x_1 and x_2 of the set of grid points. From section 2 we know that with probablity $(1 - \alpha)$ the true curve $m(x)$ lies at these points in $[\underline{c}(x_j), \overline{c}(x_j)]$, $j = 1, 2$. The exact for of $\underline{c}, \overline{c}$ is given in Theorem 2. By the mean value theorem we know that for some point $\xi \in [x_1, x_2]$: $m(x_2) - m(x_1) = m'(\xi)(x_2 - x_1)$ thus it is natural to use bounds on the first derivative. Suppose that

$$\underline{\delta} \leq m'(x) \leq \overline{\delta}, x_1 \leq x \leq x_2$$

then the two error bars can be joined with two line segments to ensure an overall upper bound between x_1 and x_2. These two lines segments are

$$x \to \overline{c}_1(x) = \overline{\delta}(x - x_1) + \overline{c}(x_1)$$
$$x \to \overline{c}_2(x) = \underline{\delta}(x - x_2) + \overline{c}(x_2).$$

In a similar way the lower bound can be constructed by

$$x \to -\underline{c}_1(x) = \underline{\delta}(x - x_1) + \underline{c}(x_1)$$
$$x \to -\underline{c}_2(x) = \overline{\delta}(x - x_2) + \underline{c}(x_2).$$

Thus the desired confidence band is

$$\underline{c}(x) = \underline{c}_1(x) \vee \underline{c}_2(x) \text{ and } \overline{c}(x) = \overline{c}_1(x) \wedge \overline{c}_2(x).$$

Obviously this set of four lines contains the true curve with probability $(1 - \alpha)$. The construction can be extended to an arbitrary set of grid points provided we have constructed error bars of simultaneous coverage probablity. Thus the confidence band is a sequence of connected hexagons.

Another approach we propose is based on bounds on the second derivative. This will lead to parabolae joining the error bars as indicated in Figure 2.

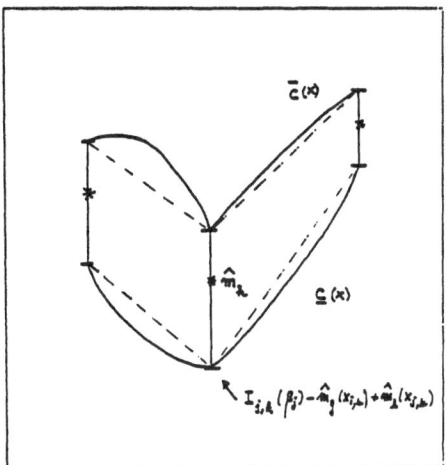

Figure 2. Bootstrap confidence bands based on bounds on the second derivative. The parabolae are different since the bound in the adjacent intervals may be different.

To get some insight into this construction consider for simplicity error bars in $x_1 = 0$, $x_2 = 1$. Let us construct the upper band. If the error bars have lengths $c(x_1)$ and $c(x_2)$ we can subtract the line joining the bars since it has second derivative equal to zero. So we may assume that the upper band $\overline{c}(x), 0 \leq x \leq 1$ passes zero at x_1 and x_2. Assume now that $|m''(x)| \leq L$ in this interval and that for some z, $m(z) \geq t > 0$. By the mean value theorem we find $\xi_j, j = 1, 2$ such that

$$m'(\xi_1) = \frac{t}{z}, m'(\xi_2) = -\frac{t}{1-z}.$$

By yet another application of the mean value theorem, there is a ξ_3 :

$$m''(\xi_3) = \frac{\frac{t}{z} - (-\frac{t}{1-z})}{\xi_2 - \xi_1}$$

$$\geq \frac{t}{z} + \frac{t}{1-z}$$

$$= \frac{t}{z(1-z)}.$$

Hence, using the bound on $m''(x)$ we have

$$t \leq L(z(1-z)).$$

This means that $m(x)$ must be below the parabola $(1-x)\overline{c}(x_1) + x\overline{c}(x_2) + L(x(1-x))$.

A different approach could be based on global bounds of the arclength of the curve between adjacent error bars. The arclength between x_1 and x_2 is $\int_{x_1}^{x_2} \sqrt{1 + (m'(x))^2}dx$, thus a bound on this quantity would give us another possibility to construct confidence bands. The geometric location of points with a fixed distance to two foci is an ellipse. The obtained confidence band would thus be a nice esthetic sequence of intersecting ellipses.

References

Cao-Abad, R. (1989). On wild bootstrap confidence intervals. Manuscript.

Cox, D. R. (1970). *Analysis of Binary Data*. New York: Chapman and Hall.

Eubank, R. (1988). *Spline Smoothing and Nonparametric Regression*. New York: M. Dekker.

Gasser, T. and Müller, H.G. (1984). Estimating regression functions and their derivatives by the kernel method. *Scandinavian Journal of Statistics*, 11, 171–185.

Härdle, W. and Bowman, A. (1988). Bootstrapping in nonparametric regression: Local adaptive smoothing and confidence bands. *Journal of the American Statistical Association*, 83, 102–110.

Härdle, W. (1990). *Applied Nonparametric Regression*. Econometric Society Monograph Series. 19. Cambridge (MA). Cambridge University Press.

Härdle, W., Huet, S. and Jolivet, E. (1990). Better bootstrap confidence intervals for regression curve estimation. Manuscript.

Härdle, W. and Mammen, E. (1990). Bootstrap methods in nonparametric regression. CORE Discussion Paper No 9049, Université Catholique de Louvain, Louvain-la-Neuve, Belgium.

Härdle, W. and Marron, J.S. (1990). Bootstrap simultaneous error bars for nonparametric regression. CORE Discussion Paper 8923, Université Catholique de Louvain, Louvain-la-Neuve, Belgium, to appear in *Annals of Statistics*.

Hall, P. (1990). On bootstrap confidence intervals in nonparametric regression. Manuscript.

Hall, P. and Titterington, M. (1986). On confidence bands in nonparametric density estimation and regression. *Journal of Multivariate Analysis*.

Knafl, G., Sacks, J., and Ylvisaker, D. (1985). Confidence bands for regression functions. *Journal of the American Statistical Association*, 80, 683–691.

APPLYING THE BOOTSTRAP TO GENERATE CONFIDENCE REGIONS IN MULTIPLE CORRESPONDENCE ANALYSIS

Monica Th. Markus, Ron A. Visser

Department of Behavioral Computer Science, Leiden University

P.O. box 9555, 2300 RB Leiden, The Netherlands[1]

1. Introduction

The bootstrap method, introduced in 1979 by Efron, promised to provide a means to solve many previously unsolved problems in statistics. Among these problems a prominent place is taken by the determination of statistical properties of complex methods for multivariate data analysis. Here we may think of multiparameter models for the exploratory analysis of multivariate data of mixed measurement level, which are usually applied in several steps (cf. Diaconis & Efron, 1983). In the last decade a large number of results on the bootstrap method have been published, most of which, however, concern univariate single parameter models. Although this attention for the relatively simple may be understood from a theoretical point of view, it is not very satisfactory for the practice of data analysis. In this paper we report on a study that may be considered as an approach from the other side. We take a complex method, i.e. multiple correspondence analysis, and try to find out what the bootstrap could contribute to data analysis. This is done mainly by Monte Carlo methods. After a short explanation of the multivariate method and the general methodology, results are reported of two Monte Carlo studies.

The general question in this study is whether bootstrap methods could generate good estimates of confidence regions. We concentrate on (1) the question whether a specific bootstrap method, the *balanced* bootstrap, could generate better results than the *plain* version of the bootstrap. In relation to this question the influence of the sample size n is considered. Additionally (2) the effect of the number of bootstrap trials T is examined.

Multiple correspondence analysis (MCA, see Gifi, 1990; Greenacre, 1985; Nishisato, 1981) belongs to a class of methods for the exploratory analysis of coherence in multivariate qualitative data. A common feature of these methods is the geometrical representation of dependencies in the data by a large number of vector-valued parameters in low dimensional Euclidian space. We may classify such methods as a form of descriptive statistics, analogous to a histogram representing properties of data in a picture that is interpreted by the statistician.

We will describe MCA in short. Suppose we have data on n independent objects which have scores on v nominal variables. The aim is to represent the objects and the categories of each variable by points in a Euclidian space. This is accomplished by the iterative quantification of the data. For dimensionality d, object scores x_{ik} ($k=1..d$; $i=1..n$) are assigned to each object i, and category quantifications y_{hjk} (category j of variable h in dimension k, $j=1..c_h$, $h=1..v$) are assigned to each category. The category quantifications y_{hjk} are defined as the average object score of all objects in category j of variable h. Vice versa, the object

[1]Netherlands organization for scientific research (NWO) is gratefully acknowledged for funding this project. This research was conducted while Drs. M.Th. Markus was supported by a PSYCHON-grant of this organization (560-267-022), awarded to Dr. R.A. Visser and Prof. W.J. Heiser.

scores are the average of the category quantifications which apply to the object. From initial values for (say) x_{ik}, optimal quantifications for x_{ik} and y_{hjk} are approximated by minimizing the loss function:

$$s(x;y)=\Sigma_h\,(\Sigma_i\,\Sigma_k\,(x_{ik}-y_{hjk})^2)$$

where y_{hjk} is the quantification of category j of variable h which *applies* to object i. The loss function can be understood as minimizing the distance between the object scores x_{ik} and the category points y_{hjk} which apply to these objects. Normalization of the object scores x_{ik} prevents trivial solutions (x_{ik}=0). To judge the quality of the quantified representation several measures of fit are defined in MCA: the *eigenvalues* λ_k represent the fit of the k^{st} dimension and the *discrimination measures* z_{hk} indicate the fit of the variable h in dimension k. Their relationship is $\lambda_k=v^{-1}\Sigma_h z_{hk}$.

Both the exploratory character and the emphasis on multivariate qualitative data, make MCA typically suited for analyses in such fields as survey research (questionnaire data) and behavioral sciences (multivariate descriptions of behavior). However, a limitation in the application of MCA is the lack of information about the stability of results under varying samples, methods of representation, etc.. Large-sample results about variances of parameter estimation are available only in certain specific situations (O'Neill, 1978). So, it is not surprising that several authors have proposed to apply the bootstrap to gain insight in a number of properties of MCA (Gifi, 1981, 1990, Ch.12; Greenacre, 1984, Ch.8, 9; see also Van der Burg & De Leeuw, 1988). In this study we concentrate on the use of the bootstrap for the estimation of non-parametric confidence regions for category quantifications and fit parameters in two dimensional solutions of MCA.

The result of a MCA solution consists of a series of statistics. For v variables, where variable h has c_h categories, and a solution in d dimensions, the result of the analysis consists of d eigenvalues, $d*v$ discrimination measures and $d*c$ category quantifications, where $c=\Sigma c_h$. For the construction of estimated two dimensional non-parametric confidence regions an algorithm is proposed that is a two dimensional analogy of the percentile method (Efron, 1982, Ch. 9). For a set of n observations T bootstrap samples (trials) of size n are generated. For each trial the values of the statistics under study are calculated. In the geometrical representation this means that a scatter of values is generated for each parameter in a d dimensional space. To generate a bivariate non-parametric confidence region, an algorithm is used based on the peeling of convex hulls of these scatter plots. Peeling here means that the points on the vertices of the hull are discarded (cf. Green, 1981). Next the convex hull of the remaining points is constructed, and so on, until a specified percentage of the points is left out. The issue of fine tuning in the approximation of a fixed percentage of outer points is discussed in detail in Markus & Visser (1990). The result of this algorithm is a polygon that is a candidate for the role of confidence region. This polygon is referred to as the x% hull.

2. Research method

The main question about the application of the bootstrap method is whether it results in good approximations of the confidence regions of the parameter values. Because few results are available about the statistical properties of MCA, we have no hope to establish this question rigorously by an analytical approach. Alternatively, we can compute finite population values. The comparison of these results with a (repeated) application of the bootstrap could indicate the validity of bootstrap results.

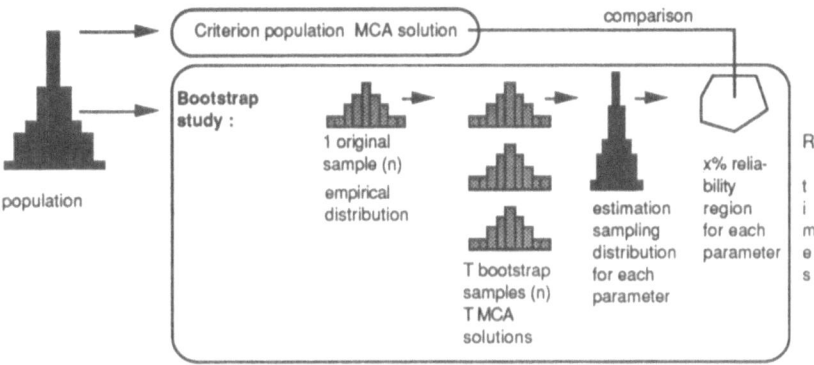

Figure 1. *General outline of the method*

Figure 1 shows a general outline of the applied method. We start with a known population. R samples of n elements are drawn from the population. For each sample the bootstrap algorithm described above, is applied. Because the population is known, the constructed R confidence regions (for each parameter) may be checked by tracing whether the population value of the parameter is included. When x% confidence regions are approximated, we expect that in the average x% of the R polygons will contain the population value. In two Monte Carlo studies this was checked for the *plain* version of the bootstrap and for the *balanced* variant of the bootstrap (Davison, Hinkley & Schechtman, 1987, Hinkley, 1988). It is conjectured that the balanced bootstrap will produce qualitatively equal results with fewer bootstrap trials. In other words, with the same amount of trials the balanced bootstrap will do better than the plain version of the bootstrap.

Two data sets are used in the Monte Carlo studies. The first set, Data I, is artificially generated. This set is based on a trivariate distribution, where each variable has three categories. The values of Cramér's V are v_{12}=.47, v_{13}=.42 and v_{23}=.46. A population of p=4000 objects was generated. A MCA solution of these objects is computed. These results can be interpreted as the values of the parameters in the population. These are the *criterion values* for the bootstrap study. The second data set, Data II contains results of a survey of attitudes with respect to nuclear and coal power, p=3983 (Midden & Verplanken, 1989). Three variables are analyzed, where the first and third variable have three categories and the second four categories.

3. Results

A valid 90% confidence region will on the average contain the true value of the parameter in 90% of the cases. We are interested to know whether the peeling algorithm described above in combination with the bootstrap method, will generate such results. If not, the question becomes: what conditions influence the type I error rate? Table 1 shows for Data I how many times (of the 100 repeated bootstrap cycles) the population value of a parameter is included in the 90% hull. We see that in case of n=50 (T=50) the balanced bootstrap outperforms the plain bootstrap. This effect is also observed in case of T=100, but the effect under this condition is weaker. The results for n=200 show no clear difference between the two forms of bootstrap methods. The comparison of two different numbers of bootstrap trials (T=50 and T=100) for the same sample size n=50, clearly shows that an increase of trials results in an improved approximation of the target value 90% under the balanced as well as under the plain bootstrap conditions (column 1 vs. 5 and 2 vs. 6).

Table 1 shows for Data II that in case of n=50 (T=50) the balanced bootstrap results are closer approximations of the value 90 than the plain bootstrap results. In this dataset the results of n=200 (T=50) confirm this result. Effects of balancing are not observed when the number of bootstrap trials is raised to 100 (column 5 vs. 6). Raising the number of bootstrap trials from T=50 to T=100 (n=50) shows the same improvement of the approximation of the value 90 as in the first study. Note there are extremely low values for the eigenvalues and the discrimination measure z_2. This appears to be the result of a poorly fitting variable, see Markus & Visser (1990) for details.

Table 1. *Results for Data I and Data II. The resulting number of cases (of the 100 repeated bootstrap cycles) in which the population criterion value is included in the 90% hull, for each parameter (the eigenvalues, discrimination measures z_h and category quantifications y_{hj}). The number of bootstrap trials, the sample size n and the bootstrap form, plain or balanced, are indicated.*

	Data I						Data II					
Trials	50	50	50	50	100	100	50	50	50	50	100	100
Sample size	50	50	200	200	50	50	50	50	200	200	50	50
Bt. form	plain	bal.	plain	bal.	plain	bal.	plain	bal.	plain	bal.	plain	bal.
Eigenvalue	76	77	73	79	88	84	0	0	0	0	0	0
z_1	76	84	80	83	88	89	88	90	95	98	92	94
z_2	84	84	68	72	89	84	0	0	33	28	0	4
z_3	74	79	75	75	84	87	82	86	96	95	90	90
y_{11}	76	76	74	72	83	80	58	51	59	58	64	69
y_{12}	67	75	75	77	83	86	57	53	57	61	68	66
y_{13}	79	77	78	76	83	91	66	70	66	71	79	78
y_{21}	74	75	74	70	85	85	75	77	77	80	84	85
y_{22}	70	68	79	75	84	86	71	77	75	75	83	79
y_{23}	74	84	76	75	80	86	67	66	61	63	73	72
y_{24}	-	-	-	-	-	-	72	74	78	78	81	80
y_{31}	73	74	71	75	81	83	62	58	67	70	69	72
y_{32}	72	75	75	73	81	83	48	51	59	55	60	62
y_{33}	78	75	75	74	88	84	64	64	79	79	75	71

4. Conclusion

The most apparent conclusion is that for both data sets the approximations of x% are systematically lower than 90%. This means that the generated regions are too small and/or biased. These results are found under the plain bootstrap condition as well as under the balanced bootstrap condition. The question is whether this could be solved by increasing the number of bootstrap trials T or an adjustment of the peeling algorithm.

The effect of an increment of bootstrap trials, from T=50 to T=100, is a clearly improved approximation of the value of 90%, under both forms of bootstrap conditions. Efron (1988) states that in case of nonparametric confidence intervals T=1000 suffices, but T=2000 is not excessive. It takes about 60 hours on a Vax 3100 workstation to compute the results for T=2000, (R=1, n=200). So, it is interesting to know whether the effect of increasing T levels off, so that values lower than these advised values may be chosen.

In both studies the balanced bootstrap results seem to outrival the results from the plain bootstrap for small sample sizes and a small number of bootstrap trials. However, with an increment of the number of bootstrap trials and the number of objects the beneficial effect of balancing diminishes. In this study we focussed on the question whether the confidence regions include the population values. It would be interesting to study also the effect of balancing on another dependent variable, namely the magnitude of the surface of the confidence region.

The phenomenon of the extremely low values for the eigenvalues and the discrimination values makes clear that in case of a low fit of categories in the population solution in combination with a weak structure, there is a possibility that the proposed procedure for estimating confidence regions deteriorates. The category quantifications do not seem to be too sensitive; the confidence regions still contain the population values in a high percentage of cases under these conditions.

References

Davison A.C., D.V. Hinkley & E. Schechtman (1987). Efficient bootstrap simulation. *Biometrika, 74*, 555-566.

Diaconis, P. & B. Efron (1983). Computer-intensive methods in statistics, *Scientific American, 17*, 96-108.

Efron, B. (1982). *The jackknife, the bootstrap and other resampling plans.* Philadelphia: SIAM.

Efron, B. (1988). Bootstrap confidence intervals: good or bad? *Psychological Bulletin, 104*, 293-296.

Gifi, A. (1990). *Nonlinear multivariate analysis.* Chichester: John Wiley & Sons.

Green, P.J. (1981). Peeling bivariate data. In: V. Barnett (ed.). *Interpreting Multivariate Data,* 3-19. New York: John Wiley & Sons.

Greenacre, M.J. (1984). *Theory and applications of correspondence analysis.* London: Academic Press.

Hinkley, D.V. (1988), Bootstrap Methods [with discussion]. *J. R. Statist. Soc,* Series B, *50(3)*, 321-337, 355-370.

Markus, M.Th. & R.A. Visser (1990). *Bootstrap methods for generating confidence regions in HOMALS; Balancing, sample size and number of trials.* Leiden: Dept. of Behavioral Comp. Science, FSW/RUL, RR-90-02.

Midden, C.J.H. & B.Verplanken (1989). *Publieksoordelen over kernenergie en kolen,* Leiden: E&M/R-89/15.

Nishisato, S. (1980). *Analysis of categorical data: dual scaling and its applications.* Toronto: University of Toronto Press.

O'Neill, M.E. (1978). Asymptotic distributions of the canonical correlations from contingency tables. *Aust. J. Statist., 20*, 75-82.

Van der Burg, E. & De Leeuw, J. (1988). Use of the multinomial jackknife and bootstrap in generalized nonlinear canonical correlation analysis. *Applied Stochastic Models and Data Analysis, 4*, 159-172.

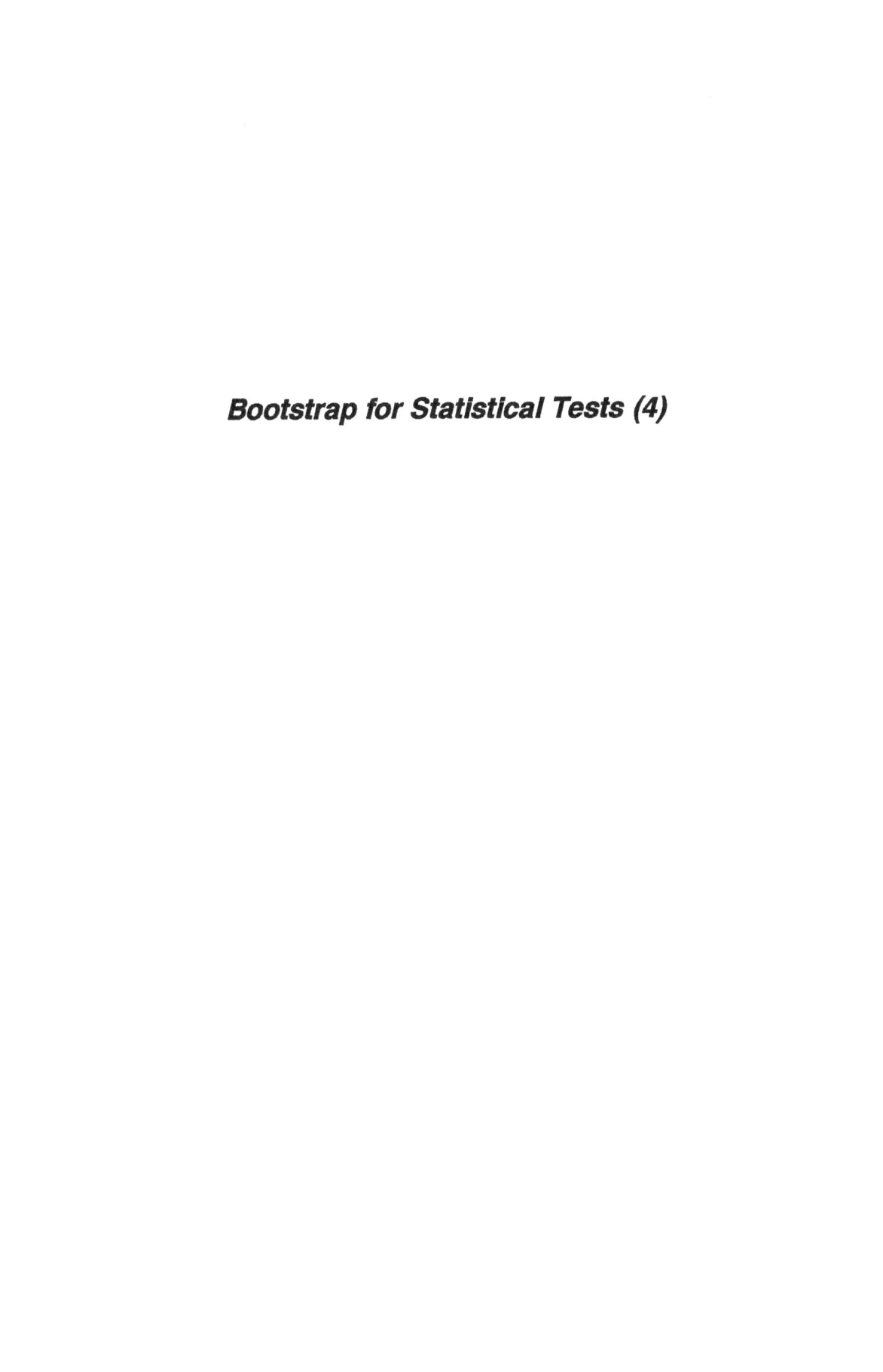

Bootstrap for Statistical Tests (4)

PERMUTATION VERSUS BOOTSTRAP SIGNIFICANCE TESTS
IN MULTIPLE REGRESSION AND ANOVA

Cajo J. F. ter Braak

Agricultural Mathematics Group

Box 100, 6700 AC Wageningen, the Netherlands

Kempthorne's (1952) formulation of the randomization test is extended to yield a permutational analog of the bootstrap significance test. In the new test, residuals of a multiple regression are permuted instead of being bootstrapped. The test is an attractive alternative for Oja's test that permutes predictors (Austr. J. Statist. 29, 91-100, 1987).

1. Introduction

Permutation tests have a long history dating back at least to Neyman (1923). As there were no computers at that time, permutation tests were not much used, but served as an additional rationale for the validity of the ANOVA F-test. Pitman (1937) and Kempthorne (1952) showed that, in simple cases, the F-test yielded an approximation to the full permutation test. Permutation tests have a simple basis for their validity: randomization in experimental design (Kempthorne 1952, Scheffé 1959) and the exchangeability assumption in observational studies. Permutation tests and related nonparametric tests based on ranks (Cox & Hinkley 1974, Lehmann 1975) exist only for simple cases and for these cases they give exact significance levels. Their application is, however, problematic both in regression (Brown & Maritz 1982, Oja 1987) and in more complex ANOVA situations, such as testing for interaction in factorial experiments. Despite the existence of Puri & Sen's (1971: section 7.7) rank test for interaction, Edgington (1987: 144) claimed that there is no valid permutation test for interaction, because "This H_0 does not refer to measurements for individual subjects, and so there is no basis for generating data permutations for alternative subject assignments, a necessity for determining significance by data permutation".

The bootstrap is relatively recent (Efron, 1979). Its main strength lies in estimating standard errors and in constructing confidence intervals (DiCiccio & Romano 1988). Bootstrap significance tests can in principle be derived from confidence intervals, but a direct approach also exists (Beran 1986, 1988, Jöckel 1990). Compared to permutation tests, the bootstrap significance test can be applied to much more complex problems, but despite the nice asymptotic properties proved by Hall & Titterington (1989) its validity in small samples is more speculative (Romano 1988, 1989).

In this paper, permutation tests are compared with bootstrap tests for use in (multivariate) regression and ANOVA. In both approaches, there is the choice what to permute or to bootstrap: the data values themselves or the residuals? In Kempthorne's (1952) formulation, the permutation test uses randomized "fixed plot errors". With errors replaced by residuals, Kempthorne's permutation test is shown to be very similar to the bootstrap test that uses residuals, the only difference being whether residuals are sampled with or without replacement. The new permutation test is briefly compared with Oja's (1987) permutation test that permutes predictors.

2. Bootstrap significance test

We consider the regression model

$$y = X\beta + Z\gamma + \varepsilon \tag{2.1}$$

where y is a random n-vector of responses on n units, X and Z are fixed known $n \times p$ and $n \times q$ matrices, the columns of which contain explanatory variables, β and γ are p- and q-vectors of unknown, fixed regression coefficients and ε contains random errors with zero mean and unknown constant variance σ^2. Our interest focuses on the effect of the variables in X on y in the presence of the covariables in Z. The parameter of interest is thus β, whereas γ and σ are nuisance parameters. In particular, we wish to test the hypothesis $\beta = \beta_0$ using the F-ratio.

Efron (1982) proposes two bootstrap schemes for regression. The one-sample bootstrap that resamples the statistical units has no permutational analog and will therefore not be considered. The other scheme resamples the observed residuals. A Monte Carlo bootstrap significance test is obtained as follows (Hall & Titterington 1989):

1. Regress y on X and Z, yielding estimates b and c for β and γ, and from these, fitted values $\hat{y} = Xb + Zc$ and residuals $e = y - \hat{y}$.
2. Calculate the F-ratio, F_{obs} say, based on y for testing the hypothesis $\beta = \beta_0$.
3. Draw a bootstrap sample e^* from e and calculate

$$\mathbf{y}^* = \mathbf{Xb} + \mathbf{Zc} + \mathbf{e}^* \tag{2.2}$$

4. Regress $\mathbf{y}^*$ on $\mathbf{X}$ and $\mathbf{Z}$, yielding estimates $\mathbf{b}^*$ and $\mathbf{c}^*$.

5. Calculate the F-ratio, F^* say, based on $\mathbf{y}^*$ for testing the hypothesis $\beta=\mathbf{b}$.

6. Repeat steps 3-5 until m bootstrap values of F^* are calculated.

7. The bootstrap estimate of the significance level is $k/(m+1)$, where k is the rank of F_{obs} among the bootstrap values of F^* when ranked from high to low.

The method works, at least in large samples, because the variability in $\mathbf{b}$ around the true value β is mimicked by the variability of $\mathbf{b}^*$ around the true value $\mathbf{b}$ in the bootstrap samples. Similarly, the variability in F_{obs} for testing $\beta=\beta_0$ is mimicked by the variability in F^* for testing $\beta=\mathbf{b}$. The mimicking is good because the test statistic is asymptotically pivotal (Hall & Titterington 1989): for normal errors, the distribution of the F-ratio does not depend on β, γ and σ and for nonnormal errors this will be asymptotically true. The procedure uses residuals under the alternative hypothesis (value of β unconstrained) to increase the power compared to a procedure using residuals from the null model (Hall & Titterington 1989).

The calculation of F^* can be simplified. Step 4 uses model (2.1) with $\mathbf{y}^*$ replacing $\mathbf{y}$. Therefore,

$$\mathbf{e}^* = \mathbf{y}^* - \hat{\mathbf{y}} = \mathbf{X}(\beta-\mathbf{b}) + \mathbf{Z}(\gamma-\mathbf{c}) + \varepsilon \tag{2.3}$$

F^* in step 5 can thus be obtained from testing the hypothesis $\beta-\mathbf{b}=\mathbf{0}$ in (2.3), i.e. by regressing $\mathbf{e}^*$ on $\mathbf{Z}$ and then adding $\mathbf{X}$ to the regression, giving residual sums of squares RSS_Z and RSS_{X+Z}, respectively. Then,

$$F^* = \{ (\text{RSS}_Z - \text{RSS}_{X+Z}) / p \} / \{ \text{RSS}_{X+Z} / (n-p-q) \} \tag{2.4}$$

3. Permutation tests

For investigating the validity of the F-test in the analysis of randomized block experiments, Kempthorne (1952: section 8.2) proposed a randomization model in which a fixed plot error is randomly assigned to treatments. He writes

"If we denote the observed yield of treatment k in block i by y_{ik}, we may write

$$y_{ik} = \mu + b_i + t_k + \sum_j \delta_{ij}^k \, e_{ij} \tag{3.1}$$

where δ_{ij}^k is equal to unity if treatment k occurs on plot j in the ith block and is zero otherwise. The random error attached to any observed yield is the whole expression $\sum_j \delta_{ij}^k \, e_{ij}$. Any particular e_{ij} is a fixed variable which we do not know. The random variable in the expression (3.1) is the term δ_{ij}^k, and its distribution is determined by the randomization procedure which is used in obtaining the particular experimental plan."

Notice that the random error in (3.1) specifies a permutation of the fixed plot errors within each block. With complete randomization, the model entails that we could equally well have observed, using our regression notation,

$$\mathbf{y}^+ = \mathbf{X}\beta + \mathbf{Z}\gamma + \varepsilon^+ \tag{3.2}$$

where ε^+ denotes a permutation of the error vector ε. By replacing the unknown β, γ and ε by their estimates, we obtain the permutational analog of (2.2)

$$\mathbf{y}^+ = \mathbf{Xb} + \mathbf{Zc} + \mathbf{e}^+ \tag{3.3}$$

yielding estimates $\mathbf{b}^+$ and $\mathbf{c}^+$. We consider this analog further in section 4.

In developing a permutation test for use in multiple regression and covariance analysis, Oja (1987) rejected (3.2) on the basis that the errors in (3.2) are unknown, but required for his test statistic. Instead, he stressed that randomization randomly assigns treatment values to units, so that the model becomes

$$\mathbf{y} = \mathbf{X}^+\beta + \mathbf{Z}\gamma + \varepsilon \tag{3.4}$$

where $\mathbf{X}^+$ is obtained by permuting the rows of the matrix $\mathbf{X}$ (see also Collins 1987). This model underlays the permutation test in the computer program CANOCO version 2.1 (ter Braak 1988a,c). For use in observational studies, this model can be motivated by assuming exchangeability of the rows of $\mathbf{X}$.

If permutations are carried out under the null hypothesis $\beta=\beta_0=\mathbf{0}$ (as is usual), then the distinction between (3.2) and (3.4) vanishes if $\gamma=\mathbf{0}$ and also if permutations are restricted within blocks, as in Kempthorne (1952). Then, it is immaterial whether $\mathbf{y}$ or the rows of $\mathbf{X}$ or the residuals $\mathbf{y}-\hat{\mathbf{y}}$ are permuted.

Model (3.4) is perfect for covariance analysis applied to randomized experiments where the covariates $\mathbf{Z}$ are measured before the random assignment of treatments to experimental units: the covariates are fixed and permutation of the rows of $\mathbf{X}$ mimics the randomization at the design stage. But, model (3.4) has a disadvantage with multiple regression: each permutation modifies the design matrix. Therefore, with each permutation the F-ratio measures the effect of another contrast (specified by the projection of X^+ on the orthocomplement of Z). To say the same more theoretically, model (3.4) violates the principle that the design is ancillary (Welch 1990), whereas model (3.3) obeys the ancillarity principle. Moreover, (3.4) does not allow for a permutation test of interaction[1], whereas (3.3) does.

4. Bootstrap versus permutation

In this section, the bootstrap (2.2) is compared with its permutational analog (3.3). In the bootstrap (2.2), residuals are resampled with replacement, whereas in permutation (3.3) the same residuals are resampled without replacement, resulting in correlation $-(n-1)^{-1}$ between the errors (Lehmann 1975: (A.41)). Because the error covariance matrices of e^* and e^+ are known, standard linear model theory (Seber 1977: section 3.7.1) can be used to obtain the expected value and the variance of b^* and b^+. The essential formulae were given by Efron (1982: example 5.6) for the bootstrap and by Cox & Hinkley (1974: example 6.2) and Lehmann (1975: (A48-49)) for permutation. The comparison is simplified if we assume that the constant vector is one of the columns in Z. Excluding regression through the origin, we obtain

$$E^*(b^*) = E^+(b^+) = b \qquad (4.1)$$

$$\mathrm{var}^*(b^*) = (1 - 1/n)\,\mathrm{var}^+(b^+) = s^2\,(\mathring{X}'\mathring{X})^{-1} \qquad (4.2)$$

where s^2 is the mean square of the residuals e and $\mathring{X}$ is obtained by projection of X on the orthocomplement of Z. Second and higher order moments of b^* and b^+ differ by $O(1/n)$. Equations (4.1-2) plus this order property form the justification for proposing the permutation model as alternative for the bootstrap model. It remains to be studied how the distributions of F^* and F^+ differ. Using permutations and bootstraps under the null hypothesis, Romano (1989) proved that the corresponding tests yield asymptotically the same critical values under quite general conditions. His results can probably be extended to our tests.

---

[1] For example, when the columns z_1 and z_2 are kept fixed, permutation of the column x_{12} which contains their elementwise product, yields an unfeasible design: x^+_{12} is not an interaction term.

5. Discussion

Traditionally, permutation is always performed under the null hypothesis and then only if the residuals are exchangeable (e.g. Puri & Sen 1971, Romano 1989, Welch 1990). Our results (4.1-2) show the validity of permutation under the alternative hypothesis, even if residuals are not exchangeable because of differences in their variances. It is thus not needed to studentize residuals. If, however, the variance of ε is heterogeneous, this knowledge can be used to standardize the residuals (Hinkley 1988: 331). In the program CANOCO version 3.1 (ter Braak 1988a,c) this device is used to obtain permutation tests for use in partial canonical correspondence analysis (ter Braak 1988b).

Freedman & Lane (1983) used a model similar to (3.3). Their proposal differs from (3.3) in that e is estimated under the null hypothesis $\beta=\beta_0$. But, the expectation and variance of b^+ have the classical form (4.1-2) only when using e from the alternative hypothesis. By contrast, in developing a permutation test for generalized linear models, Gail et al (1988) used Oja's (1987) approach (3.4) of permuting the treatment variables.

In the traditional permutation tests, the test statistic can often be simplified, e.g. from an F-ratio to a sum of squares (Edgington 1987). Using (3.3), there is no such simplication. Hall & Titterington (1989) show the gain in level accuracy by using an asymptotically pivotal statistic such as the F-ratio.

If randomization is used to obtain the data, reference to an underlying population model can be avoided by using the permutation approach (cf Pitman 1937, Cox & Hinkley 1974) instead of the bootstrap. In principle, the randomization model, whence permutation, also forms a basis for constructing confidence regions.

In practice, bootstrap and permutation are often carried out by Monte Carlo methods. Balanced bootstraps and other variance reduction techniques may then help to increase the power when only a limited number of bootstrap samples can be made (Hinkley 1988). Permutation may have some advantage here because there is maximum balance in each random permutation.

Notice that the set of all permutations is a proper subset of the set of all bootstrap samples. As a result, permutation is sometimes not applicable, e.g. in one-sample problems where all permutations yield the same mean value. For the rest, bootstrap and permutation tests are just different but related tests.

6. References

Beran, R. (1986). Simulated power functions. *Ann. Statist.* **14**, 151-173.

Beran, R. (1988). Prepivoting test statistics: a bootstrap view of asymptotic refinements. *J. Amer. Statist.* **83,** 687-697.

Brown, B. M. and Maritz, J. S. (1982). Distribution-free methods in regression. *Austral. J. Statist.* **24,** 318-331.

Collins, M. F. (1987). A permutation test for planar regression. *Austral. J. Statist.* **29,** 303-308.

Cox, D. R. and Hinkley, D. V. (1974). *Theoretical Statistics.* London: Chapman and Hall.

DiCiccio, T. J. and Romano, J. P. (1989). A review of bootstrap confidence intervals. *J. R. Statist. Soc. B* **50,** 338-354.

Edgington, E. S. (1987). *Randomization Tests. 2nd Ed.* New York: Dekker.

Efron, B. (1979). Bootstrap methods: another look at the jackknife. *Ann. Statist.* **7,** 1-26.

Efron, B. (1982). *The Jackknife, the Bootstrap and other Resampling Plans.* Philadelphia: SIAM.

Freedman, D. A. and Lane, D. (1983). A nonstochastic interpretation of reported significance levels. *J. Bus. Econ. Statist.* **1,** 292-298.

Gail, M. H., Tan, W. Y. and Piantadosi, S. (1988). Tests for no treatment effect in randomized clinical trials. *Biometrika* **75,** 57-64.

Hall, P. and Titterington, D. M. (1989). The effect of simulation order on level accuracy and power of Monte Carlo tests. *J. R. Statist. Soc. B* **51,** 459-467.

Hinkley, D. V. (1988). Bootstap methods. *J. R. Statist. Soc. B* **50,** 321-337.

Jöckel, K.-H. (1990). Monte Carlo techniques and hypothesis testing. In *Irst Int. conf. Statistical computing, Cesme, Izmir 1987,* E. J. Dudewisz (ed.).

Kempthorne, O. (1952). *The Design and Analysis of Experiments.* New York: Wiley.

Lehmann, E. L. (1975). *Nonparametrics: Statistical Methods Based on Ranks.* San Francisco: Holden-Day.

Neyman, J. (1923). On the application of probability theory to agricultural experiments: principles (in Polish with German summary). *Roczniki Nauk Rolniczch* **10,** 21-51.

Oja, H. (1987). On permutation tests in multiple regression and analysis of covariance analysis problems. *Austral. J. Statist.* **29,** 91-100.

Pitman, E. J. G. (1937). Significance tests which may be applied to samples from any populations. III The analysis of variance test. *Biometrika* **29,** 322-335.

Puri, M. L. and Sen, P. K. (1971). *Nonparametric Methods in Multivariate Analysis.* New York: Wiley.

Romano, J. P. (1988). A bootstrap revival of some nonparametric distance tests. *J. Amer. Statist.* **83,** 698-708.

Romano, J. P. (1989). Bootstrap and randomization tests of some nonparametric hypotheses. *Ann. Statist.* **17,** 141-159.

Scheffé, H. (1959). *The Analysis of Variance.* New York: Wiley.

Seber, G. A. F. (1977). *Linear Regression Analysis.* New York: Wiley.

ter Braak, C. J. F. (1988a). CANOCO - an extension of DECORANA to analyze species-environment relationships. *Vegetatio* **75,** 159-160.

ter Braak, C. J. F. (1988b). Partial canonical correspondence analysis. In *Classification and Related Methods of Data Analysis,* H. H. Bock (ed.), 551-558. Amsterdam: North-Holland.

ter Braak, C. J. F. (1988c). *CANOCO - a FORTRAN program for canonical community ordination by [partial] [detrended] [canonical] correspondence analysis, principal components analysis and redundancy analysis (version 2.1). Report LWA-88-02.* Wageningen: Agricultural Mathematics Group.

Welch, W. J. (1990). Construction of permutation tests. *J. Amer. Statist.* **85,** 693-698.

NONPARAMETRIC BOOTSTRAP TESTS: SOME APPLICATIONS

Jörg Breitung

Institut für Quantitative Wirtschaftsforschung, Universität Hannover

Wunstorfer Str. 14, D-3000 Hannover 91

1. Introduction

In a series of papers Beran (1984, 1986, 1988) proposed bootstrap techniques for hypothesis testing. These tests are concerned with the following situation. Let $\{X_1, X_2, ..., X_n\}$ be an i.i.d. sample of n random variables with distribution function F and the parameter $\theta(F)$ which is a real-valued functional statistic to be tested for $H_0 : \theta(F) = \theta_0$. To test this hypothesis Beran proposes two alternative approaches. The first one which is called the *test statistic approach* approximates the exact critical region of the test by the percentiles of its bootstrap distribution, where the unknown distribution of the sample is replaced by its empirical distribution.

At the *confidence region approach* a confidence region for the estimated parameter is constructed, and the null hypothesis is rejected if θ_0 lies outside the confidence interval. Whenever the parameters of interest can be transformed to a normal pivot, an appropriate critical region can be derived, without knowledge of the exact functional form of the transformation.

In this paper we sketch the nonparametric bootstrap approach and compare the small sample properties of the bootstrap test with the conventional parametric alternatives based on normality assumptions.

2. The test statistic approach

Let $T_n = T_n(X_1, \ldots, X_n)$ be a test statistic for the null hypothesis $H_0 : \theta = \theta_0$ vs. $H_1 : \theta \neq \theta_0$ and $F_T(\xi; \theta_0) = Prob(T_n < \xi)$ is the distribution of T_n under H_0. The critical region of the test is given by

$$d_L(\frac{\alpha}{2}; \theta_0, F) = F_T^{-1}(\frac{\alpha}{2}; \theta_0)$$

(1)

$$d_U(1 - \frac{\alpha}{2}; \theta_0, F) = F_T^{-1}(1 - \frac{\alpha}{2}; \theta_0) \quad ,$$

where α is the significance level of the test. However, in most applications $F_T(\cdot)$ is unknown and therefore have to be replaced by estimates

$$d_L^*(\frac{\alpha}{2}; \theta_0, \hat{F}) = F_T^{*-1}(\frac{\alpha}{2}; \theta_0)$$

(2)

$$d_U^*(1 - \frac{\alpha}{2}; \theta_0, \hat{F}) = F_T^{*-1}(1 - \frac{\alpha}{2}; \theta_0) \quad ,$$

where $F_T^*(\cdot)$ is the bootstrap distribution of T_n based on the empirical distribution of X given H_0 is true.

A nonparametric approach is to approximate the unknown distribution of X by the empirical distribution $\hat{F}(\xi) = \frac{1}{n} \sum I(X_i \leq \xi)$, where $I(A)$ is an indicator function which is one if A is true. Given H_0, the empirical distribution approaches the true distribution of X under H_0 as $n \to \infty$ and therefore the test is asymptotically of correct size.

3. The confidence region approach

Let $R_n(\mathbf{X}, \theta)$ be a pivot for θ, where $\mathbf{X} = (X_1, X_2, \ldots, X_n)$ and $F_R(\xi; \theta)$ is the corresponding c.d.f. of R_n. The critical values of the confidence region approach are

(3)
$$c_L(\frac{\alpha}{2}; \hat{\theta}, F) = F_R^{-1}(\frac{\alpha}{2}; \hat{\theta})$$
$$c_U(1 - \frac{\alpha}{2}; \hat{\theta}, F) = F_R^{-1}(1 - \frac{\alpha}{2}; \hat{\theta}) \quad .$$

Replacing F_R by its bootstrap approximation yields

(4)
$$c_L^*(\frac{\alpha}{2}; \hat{\theta}, \hat{F}) = F_R^{*-1}(\frac{\alpha}{2}; \hat{\theta})$$
$$c_U^*(1 - \frac{\alpha}{2}; \hat{\theta}, \hat{F}) = F_R^{*-1}(1 - \frac{\alpha}{2}; \hat{\theta}) \quad .$$

However it is sometimes difficult to find the pivot $R(\mathbf{X}, \theta)$ and for some parametric families no such pivot exists. Furthermore, in nonparametric cases pivots should be constructed which hold for all possible distributions. Efron (1979, Remark B) considered the class of estimators $\hat{\theta}$ for which a monotonic increasing function $g(\cdot)$ exists, such that the transformed quantities $\phi = g(\theta)$ and $\hat{\phi} = g(\hat{\theta})$ satisfy

(5)
$$(\hat{\phi} - \phi)/\tau \quad \sim \quad N(0, 1) \quad ,$$

where τ is the standard error of $\hat{\phi}$. Since the percentil method for constructing bootstrap confidence intervals is transformation invariant, the normalization to the pivot (5) is automatically incorporated. Efron (1982, 1987) extends the results to

(6)
$$BC : \quad (\hat{\phi} - \phi)/\tau \quad \sim \quad N(-z_0, 1)$$

and

(7)
$$BC_a : \quad (\hat{\phi} - \phi)/\tau \quad \sim \quad N(-z_0(1 + a\phi), (1 + a\phi)^2) \quad ,$$

where z_0 and a are suitable chosen constants.

4. Testing the mean

Testing the sample $X_1, X_2, \ldots, X_n \overset{i.i.d.}{\sim} F$ for $H_0 : E(X_i) = \mu_0$ the test statistic $T_n = \sqrt{n}(\bar{X} - \mu_0)/\hat{\sigma}$ is usually applied, where $\bar{X} = \frac{1}{n}\sum X_i$ and $\hat{\sigma}^2 = \frac{1}{n-1}\sum(X_i - \bar{X})^2$. To approximate the distribution of T_n observations of T_n^* were computed by drawing n elements from the realized sample $\{x_1, x_2, \ldots, x_n\}$ and calculating

$$(8) \qquad T_n^* = \sqrt{n}\frac{\bar{X}^* - \bar{X}}{\hat{\sigma}^*} \quad ,$$

where $\bar{X}^*$ and $\hat{\sigma}^*$ are defined similar to $\bar{X}$ and $\hat{\sigma}$. The histogram of m realizations of T_n^* approaches the distribution of T_n under H_0 for large n and m.

Let $F_T^*(\xi) = \frac{1}{m}\sum I(T_{(i)}^* \leq \xi)$ be the bootstrap distribution of T_n, where $T_{(i)}^*$ indicates the ith Monte Carlo realization of T_n^*, then the critical values result from

$$
\begin{aligned}
d_L^* &= F_T^{*-1}(\frac{\alpha}{2}) \\
d_U^* &= F_T^{*-1}(1 - \frac{\alpha}{2}).
\end{aligned}
$$
(9)

The null hypothesis is rejected if T_n lies outside the interval $[d_L^*, d_U^*]$.

For the confidence region approach we assume that there exists a transformation $g(\cdot)$ for $\bar{X}$ and μ which provides a normal pivotal quantity. Again m sets of realizations from $\{X_1^*, X_2^*, \ldots, X_n^*\}$ are generated to compute the bootstrap distribution of $\bar{X}$

$$(10) \qquad F_{\bar{X}}^*(\xi) = \frac{1}{m}\sum_{i=1}^{m} I(\bar{X}_{(i)}^* \leq \xi) \quad .$$

The critical region results from inverting the bootstrap distribution at $\frac{\alpha}{2}$ and $1 - \frac{\alpha}{2}$

$$
\begin{aligned}
c_L^* &= F_{\bar{X}}^{*-1}(\frac{\alpha}{2}) \\
c_U^* &= F_{\bar{X}}^{*-1}(1 - \frac{\alpha}{2}).
\end{aligned}
$$
(11)

The test rejects H_0 if μ_0 lies outside this interval.

If the normal pivot is of the form (6) a correction is necessary. In this case the critical value results from

$$(12) \quad BC : \qquad c_L^{z_0} = F_{\bar{X}}^{*-1}(\Phi\{2z_0 + z^{(\alpha/2)}\})$$

$$c_U^{z_0} = F_{\bar{X}}^{*-1}(\Phi\{2z_0 - z^{(\alpha/2)}\}).$$

For the critical values of the BC_a approach according to (7) see Efron (1987).

The difference of the two approaches in this application is that for the test statistic approach a studentized statistic is bootstrapped, whereas for the confidence region approach the bootstrapped statistic is not studentized. From bootstrapping studentized statistics, however, a higher accuracy is expected (Abramovith and Singh 1985, Rothe 1986). Table 1 presents some results of a Monte Carlo study comparing the empirical size of the one-sided bootstrap test (T_n^*) with the ordinary t-test (T_n) according to a nominal size of $\alpha = 0.10$ and in table 2 we give the empirical sizes of the percentile method (R^*) and the bias corrected intervals (R_{BC}^*).

Table 1: Empirical sizes for the test of H_0: $\mu \geq 1$ applying the test statistic approach

n	normal		uniform		exponential	
	T_n	T_n^*	T_n	T_n^*	T_n	T_n^*
20	0.097	0.098	0.097	0.089	0.157	0.113
30	0.103	0.106	0.106	0.091	0.150	0.107
50	0.103	0.101	0.096	0.096	0.143	0.110
100	0.101	0.103	0.107	0.108	0.116	0.096

Rejection frequencies under H_0 of 2000 Monte Carlo iterations with 1000 bootstrap replications each. Standard errors are about 0.007. The nominal size is α=0.10.

Table 2: Empirical sizes for the test of H_0: $\mu \geq 1$ applying the confidence region approach

n	normal		uniform		exponential	
	R^*	R_{BC}^*	R^*	R_{BC}^*	R^*	R_{BC}^*
20	0.116	0.113	0.120	0.116	0.078	0.091
30	0.105	0.105	0.121	0.118	0.074	0.092
50	0.108	0.112	0.102	0.105	0.079	0.094
100	0.096	0.098	0.106	0.107	0.087	0.099

Rejection frequencies under H_0 of 2000 Monte Carlo iterations with 1000 bootstrap replications each. Standard errors are about 0.007. The nominal size is α=0.10.

5. Testing the variance

In this section we consider bootstrap tests of H_0 : $E(X_i - \mu)^2 = \sigma_0^2$ from a sample $X_1, X_2, \ldots, X_n \overset{i.i.d.}{\sim} F$ where we suppose $E(X_i) = \mu < \infty$ and $E(X_i - \mu)^2 = \sigma^2 < \infty$. The LR-test statistic

$$(13) \qquad T_n = n \left[\frac{\hat{\sigma}^2}{\sigma_0^2} - \ln \left(\frac{\hat{\sigma}^2}{\sigma_0^2} \right) - 1 \right] = f \left(\frac{\hat{\sigma}^2}{\sigma_0^2} \right)$$

is asymptotically distributed as χ_1^2.

Since the bootstrap procedure is transformation invariant, we can consider the statistic $\tilde{T}_n = \hat{\sigma}^2/\sigma_0^2$ which is a monotonic function of T_n. To derive the bootstrap distribution of $\tilde{T}_n$ under H_0 the simulated data X^* have to be corrected for their scale parameter, such that

$$(14) \qquad \qquad \text{Var}_*(\lambda_0 X_1^*) = \sigma_0^2 \quad ,$$

where $\lambda_0 = \sqrt{\frac{n}{n-1}\sigma_0^2/\hat{\sigma}^2}$. Hence

$$(15) \qquad \qquad \tilde{T}_n^* = \frac{\frac{1}{n-1}\sum_{i=1}^n(\lambda_0 X_i^* - \lambda_0 \bar{X}^*)^2}{\sigma_0^2} = \frac{n}{n-1}\frac{\sigma^{*2}}{\hat{\sigma}^2} \quad .$$

The critical values were found by inverting the bootstrap c.d.f. of $\tilde{T}_n$ at $\frac{\alpha}{2}$ and $1 - \frac{\alpha}{2}$. In table 3 the results of a Monte Carlo simulation are presented to compare the performance of the bootstrap test $(\tilde{T}_n^*)$ with the asymptotic test (T_n).

Table 3: Empirical sizes for the test of H_0: $\sigma^2 \geq 1$ applying the test statistic approach

n	normal		uniform		exponential	
	T_n	$\tilde{T}_n^*$	T_n	$\tilde{T}_n^*$	T_n	$\tilde{T}_n^*$
20	0.095	0.118	0.141	0.069	0.053	0.253
30	0.101	0.113	0.148	0.061	0.057	0.225
50	0.097	0.118	0.157	0.074	0.049	0.195
100	0.107	0.113	0.153	0.075	0.052	0.165

Rejection frequencies under H_0 of 2000 Monte Carlo iterations with 1000 bootstrap replications each. Standard errors are about 0.007. The nominal size is α=0.10.

To apply the confidence region approach the bootstrap distribution of $\hat{\sigma}^2$ is computed from m observations of $\sigma^{*2} = \frac{1}{n-1}\sum_{i=1}^n(X_i^* - \bar{X}^*)^2$, i. e.

$$(16) \qquad \qquad F_{\hat{\sigma}^2}^*(\xi) = \frac{1}{m}\sum_{i=1}^m I(\sigma_{(i)}^{*2} \leq \xi)$$

where $\sigma_{(i)}^{*2}$ indicates the outcome of the i'th resampling experiment.

Usually no transformation $g(\cdot)$ exists which leads to a normal pivot of the form $N(0,1)$. Therefore, corrected confidence intervals according to (12) are necessary to improve the test. The empirical sizes of the percentil method (R^*) and the BC-method (R_{BC}^*) are presented in table 4. Obviously the bias corrected intervals perform much better, due to the skewness of the distribution of $\hat{\sigma}^2$.

Table 4: Empirical sizes for the test of H_0: $\sigma^2 \geq 1$ applying the confidence region approach

	normal		uniform		exponential	
n	R^*	R^*_{BC}	R^*	R^*_{BC}	R^*	R^*_{BC}
20	0.050	0.127	0.034	0.143	0.034	0.075
30	0.047	0.110	0.053	0.133	0.030	0.071
50	0.046	0.097	0.071	0.135	0.037	0.073
100	0.067	0.109	0.079	0.117	0.042	0.072

Rejection frequencies under H_0 of 2000 Monte Carlo iterations with 1000 bootstrap replications each. Standard errors are about 0.007. The nominal size is $\alpha=0.10$.

6. Further Applications

In the extended version of this paper we apply bootstap tests in regression models and to test for biased estimators. Furthermore nested bootstrap procedures are considered.

References

Abramovitch, L. and Singh, K. (1985), Edgeworth corrected pivotal statistics and the bootstrap, Ann. Statist., 13, 116–132.

Beran, R. J. (1984), Bootstrap methods in statistics, Jber. der Dt. Math. Verein., 86, 14–30.

Beran, R. J. (1986), Simulated power functions, Ann. Statist., 14, 151–173.

Beran, R. J. (1988) Prepevoting test statistics: A bootstrap view of asymptotic refinements, J. Amer. Statis. Assoc. 83, 687–697.

Efron, B. (1979), Bootstrap methods: Another look at the jackknife. Ann. Statist., 7, 1–26.

Efron, B. (1982), The jackknife, the bootstrap, and other resampling plans, CBMS Monograph #38, Society for Industrial and Applied Mathematics, Philadelphia.

Efron, B. (1987), Better bootstrap confidence interval, J. Amer. Statist. Assoc., 82, 171–185.

Rothe, G. (1986), Some remarks on bootstrap techniques for constructing confidence intervals, Statistische Hefte, 27, 165–172.

A CLASS OF COMBINATIONS OF DEPENDENT TESTS BY A RESAMPLING PROCEDURE

A . Pallini and F. Pesarin

Dept. of Statistical Sciences, University of Padua

Via S. Francesco 33, 35121 Padova (Italy).

1. Introduction.

A number of statistical hypotheses could be tested by combining $k > 2$ test statistics in a single statistic. e.g. hypotheses for the equality of several means and variances, or more generally for different effects jointly relevant for the analysis. In this paper, we show how resampling techniques, based on permutations of the data, may be conveniently used to combine the k test statistics, when they are characterized by an unknown dependence structure. These techniques have recently been interpreted (Hinkley (1989), Pesarin (1989) and Romano (1989)) as conditional bootstrap methods.

More precisely, let X be an observable random variable (r.v.) with values in a space S_n and probability distribution P_n defined on a σ-algebra W_n of subsets $W \subset S_n$. Given a random sample $\mathbf{x}_n = (x_1, \ldots, x_n)$ of n observations, we wish to test the null hypothesis H_0 that the unknown probability distribution generating the data belongs to a subclass $\Omega_0 \subset \Omega$, i.e. $H_0 : P_n = P_n^0$, against the alternative $H_1 : P_n = P_n^1$, where $P_n^0 \in \Omega_0$, $P_n^1 \in \Omega_1$ and typically $\Omega_0 \cup \Omega_1 = \Omega$. A collection $\mathcal{T} = \{T_{ni}, i \in I\}$ of real-valued test statistics appropriate for this testing problem is available. The index set $I = \{1, \ldots, k\}$ is finite and does not vary with n. The statistics $\{T_{ni}, i \in I\}$ are all based on $\mathbf{x}_n$, i.e. $\{T_{ni}(\mathbf{x}_n), i \in I\}$, W_n-measurable and assumed to be dependent. We wish to study the asymptotic behaviour of a class of permutation combinations $T_n^c(\mathbf{x}_n)$ of the test statistics $\{T_{ni}(\mathbf{x}_n), i \in I\}$.

2. A class of combinations of dependent tests.

In order to choose a suitable test statistic $T_n^c(\mathbf{x}_n)$, let us consider a real-valued and W_n-measurable combining function defined as

$$T_n^c(\mathbf{x}_n) = \sum_{i \in I} \psi_i [T_{ni}(\mathbf{x}_n)] \tag{1}$$

where the scores $\psi_i = \psi_i(t)$, $i \in I$, are unbounded and non-decreasing functions of t, for every $t \in \mathbb{R}^1$.

Let the test $\phi_n^c(\mathbf{x}_n)$ defined by

$$\phi_n^c(\mathbf{x}_n) = \begin{cases} 1 & \text{if } T_n^c(\mathbf{x}_n) > \gamma_n, \\ a_n^c(\mathbf{x}_n) & \text{if } T_n^c(\mathbf{x}_n) = \gamma_n, \\ 0 & \text{if } T_n^c(\mathbf{x}_n) < \gamma_n. \end{cases} \tag{2}$$

where the constants $0 \le a_n^c(\mathbf{x}) \le 1$ and γ_n are such that $\phi_n^c(\mathbf{x})$ is of size α, $0 < \alpha < 1$, for testing H_0; without loss of generality, the test $\phi_n^c(\mathbf{x}_n)$ relates to one-sided alternatives.

The class $\mathcal{C}$ of tests $\phi_n^c(\mathbf{x}_n)$ based on possible specifications of (1) is quite general and contains some well known combining procedures, the most important of which are those of Fisher and Liptak (cf. Folks (1984)). If the test statistics $T_{ni}(\mathbf{x}_n)$, $i \in I$, are independent, the Fisher test is Bahadur optimal (Littell-Folks (1973)) in the class $\mathcal{C}$. Pesarin (1989) has investigated conditions under which the test $\phi_n^c(\mathbf{x}_n)$ is consistent and unbiased, without any assumption about the independence of the $T_{ni}(\mathbf{x}_n)$, $i \in I$.

It should be noted that it is very difficult to obtain the null distribution function (d.f) of $T_n^c(\mathbf{x}_n)$ and hence to apply $\phi_n^c(\mathbf{x}_n)$ (cf. Berk-Jones (1978)), if the test statistics $T_{ni}(\mathbf{x}_n)$, $i \in I$, are dependent. A possible solution of this problem is to use a permutation counterpart of $\phi_n^c(\mathbf{x}_n)$, i.e. a conditional test, when it may be applied.

3. Combinations based on permutational principle.

Let G_n be a finite group of M_n transformations of S_n onto itself, which also maps W_n onto itself. Thus, for every $g \in G_n$, every $\mathbf{x}_n \in S_n$ and every $W \in W_n$, the point $g\mathbf{x}_n$ and the set gW are still in S_n and in W_n respectively. We assume that the null hypothesis H_0 implies that the distribution of $\mathbf{x}_n$ is invariant under transformations in G_n, so that for every g in G_n, $g\mathbf{x}_n$ and $\mathbf{x}_n$ have the same distribution.

For every $\mathbf{x}_n$ in S_n let $T_{n(1)}^c(\mathbf{x}_n) \le \ldots \le T_{n(M_n)}^c(\mathbf{x}_n)$ be the ordered values of $T_n^c(g\mathbf{x}_n)$, for all $g \in G_n$. Given α, let $m_n = M_n - [M_n \alpha]$, where [t] denotes the largest integer less than or equal to t. Let $M_n^+(\mathbf{x}_n)$ and $M_n^0(\mathbf{x}_n)$ the numbers of values of $T_{n(j)}^c(g\mathbf{x}_n)$, $j=1,\ldots,M_n$, which are greater than $T_{n(m_n)}^c(\mathbf{x}_n)$ and equal to $T_{n(m_n)}^c(\mathbf{x}_n)$ respectively.

Let $\phi^*(\mathbf{x}_n)$ defined by

$$\phi_n^*(x_n) = \begin{cases} 1 & \text{if } T_n^c(x_n) > T_{n(m_n)}^c(x_n), \\ a_n^*(x_n) & \text{if } T_n^c(x_n) = T_{n(m_n)}^c(x_n), \\ 0 & \text{if } T_n^c(x_n) < T_{n(m_n)}^c(x_n). \end{cases} \tag{3}$$

where $a_n^*(x_n) = [M_n \alpha - M_n^+(x_n)] / M_n^0(x_n)$. Since P_n^0 is invariant under G_n, $T_n^c(x_n)$ is equally likely to be any of the M_n values $T_n^c(gx_n)$, with $g \in G_n$. This implies (cf. Hoeffding (1952)) that $\phi_n^*(x_n)$ is similar of size α, i.e. $E_{P_n^0}[\phi_n^*(x_n)] = \alpha$, for testing H_0. Let $G_n x_n = \{gx_n | g \in G_n\}$ be the G_n-orbit of x_n in S_n. Define $M_n \cdot F_n(y | G_n x_n)$ to be the number of $g \in G_n$ for which $T_n^c(gx_n) \le y$; for x_n fixed, $F_n(y | G_n x_n)$ is the conditional d.f. of $T_n^c(x')$, given that x_n' is in $G_n x_n$.

A number of studies on test of the form (3) has been published; see Puri-Sen (1971) and Bell-Sen (1984) for a review. In particular, it is important to note that for such conditional tests are characterized by critical values which are r.v.'s; this makes an exact evaluation of the power very difficult. However, it can be shown from Hoeffding (1952), that the power functions of $\phi_n^*(x_n)$ and $\phi_n^c(x_n)$ are appropriately close as $n \to \infty$.

Assume that the objects considered, i.e. x_n, S_n, P_n, $\phi_n^c(x_n)$, $T_n^c(x_n)$, $\phi_n^*(x_n)$, etc. are defined on infinite sequence of positive integers and that $M_n \to \infty$ as $n \to \infty$. If α is a fixed constant, $M_n \to 1-\alpha$ as $n \to \infty$.

For convenience, let us replace $T_n^c(x_n)$ in (3) of by a test statistic, say $T_n^e(x_n)$, such that for any two elements g and g' in G_n, the differences $T_n^c(gx_n) - T_n^c(g'x_n)$ and $T_n^e(gx_n) - T_n^e(g'x_n)$ have the same sign; $T_n^c(x_n)$ and $T_n^e(x_n)$ are called equivalent (cf. Hoeffding (1952)), in the sense that they have the same conditional d.f. $F_n(y | G_n x_n)$. Let us define the new parametric analogue of (3) as

$$\phi_n^e(x_n) = \begin{cases} 1 & \text{if } T_n^e(x_n) > \lambda_n, \\ a_n^e(x_n) & \text{if } T_n^e(x_n) = \lambda_n, \\ 0 & \text{if } T_n^e(x_n) < \lambda_n. \end{cases} \tag{4}$$

where the constants $0 \le a_n^e(x) \le 1$ and γ_n are such that $\phi_n^e(x)$ is of size α. For every element of the sequence $\{P_n\}$, let $F_n(y, P_n) = \Pr\{T_n^e(x_n) \le y\}$ be the d.f of $T_n^0(x_n)$.

Under the following conditions (Hoeffding (1952)), the power functions of $\phi_n^*(x_n)$ and $\phi_n^e(x_n)$ converge in probability ($\xrightarrow{p}$) to the same limit as $n \to \infty$.

Condition A: $F_n(y | G_n x_n) \xrightarrow{p} F(y)$ as $n \to \infty$, for every y at which $F(y)$ is continuous, where $F(y)$ is a d.f. and the equation $F(y) = 1-\alpha$ has a unique *solution at y= λ.*

Condition B: there exists a function Z(y), continuous at y= λ, such that for every y at which Z(y) is continuous $F_n(y|x_n) \overset{p}{\longrightarrow} Z(y)$.

Note that A and B imply $E_P \phi^*(x_n) \overset{p}{\longrightarrow} 1 - Z(y)$ (Hoeffding (1952)); under A the critical value $T^e_{n(m_n)}(x_n) \overset{p}{\longrightarrow} \lambda$, i.e. to a fixed constant. For every element of $\{P^0_n\}$, $F_n(y, P_n) \overset{p}{\longrightarrow} F(y)$, i.e. $T^e_n(x_n)$ and $T^e_n(G_n x_n)$ have the same asymptotic null d.f.; we also have that $\lambda_n \longrightarrow \lambda$. B finally implies that $E_{P_n} \phi^c(x_n) \overset{p}{\longrightarrow} 1 - Z(y)$. Let $\mathcal{D}(\lambda)$ the class of all sequences $\{P_n\}$ for which A, with λ fixed, and B, for some Z(y), are satisfied If $\mathcal{D}(\lambda)$ contains $\{P^0_n\}$, for every $\{P_n\}$ in $\mathcal{D}(\lambda)$, the conditional test $\phi^*_n(x_n)$ is asymptotically as powerful as $\phi^e_n(x_n)$ with respect to $\mathcal{D}(\lambda)$.

To obtain the validity of A and B, suppose to consider test statistic of the form $T_{ni}(x_n) = \Sigma_{1 \leq j \leq n} b_{nij} x_j$, i∈I, where coefficients b_{nij} are not all equal and may depend on n. A test statistic equivalent to $T^c_n(x_n)$, that may be obtained from (1) with $\psi_i(t) = t$, $t \in \mathbb{R}^1$, is

$$T^e_n(x_n) = \sum_{1 \leq j \leq n} c_{nj} x_j \qquad (5)$$

where c_{nj} is an increasing function of $\Sigma_{i \in I} b_{nij}$, such that $\Sigma_{1 \leq j \leq n} (c_{nj} - \bar{c}_n)^2$, with $\bar{c} = (1/n)\Sigma_{1 \leq j \leq n} c_{nj}$, and $(1/n) \underset{1 \leq j \leq n}{\max} (c_{nj} - \bar{c}_n)^2 \to 0$ as n→ ∞. From Dwass (1955), Theorem 3.1, it follows that, if the r.v's x_j, j= 1,...,n, have finite variances, (5) is normally distributed, when standardized. Thus, A and B are fulfilled. Since (5) is an increasing function of (1), $\phi^e_n(x_n)$ and $\phi^c_n(x_n)\phi$ have the same power function; hence, $\phi^*_n(x_n)$ and $\phi^c_n(x_n)$ *have the same limit power for testing* H_0 *against the alternatives considered.*

Remark 1. Instead of (5), it is possible to consider any test statistic, which can be suitably approximated by a linear function of the elements of x_n.

Remark 2. Note that, any test $\phi^*_n(x_n)$ defined by (5), which is nonparametric wuth respect to the unknown dependences among $T_{ni}(x_n)$, i∈I, is asymptotically as powerful as any parametric test $\phi^c_n(x_n)$ defined by (1).

Remark 3. If a test, belonging to class $\mathcal{C}$, has a "best" power function (in some sense), then $\phi^*_n(x_n)$ defined by (5), has asymptotically the same optimalities.

4. Stochastic approximations.

If the number of elements in G_n is too large, the conditional d.f. $F_n(y|G_n\underline{x}_n)$ may be very hard to evaluate. In this case, we sample $g_1,\ldots,g_r$ from G_n, with or without replacement, and apporoximate (cf. Dwass (1957) and Vadiveloo (1983)) the conditional distribution of $T_n^e(\underline{x}_n)$, by a distribution assigning equal mass to the values $T_n^e(g_h\underline{x}_n)$, $1 \leq h \leq r$. Let $\tilde{F}_n(y|G_n\underline{x}_n)$ be the resulting empirical conditional d.f.. such form of stochastic approximation, based on monte Carlo simulation, does not affect the desired level α of the test (3), because (cf. Romano (1989)) the random vector $\{T_n^c(\underline{x}_n),$ $T_n^c(g_1\underline{x}_n),\ldots,T_n^c(g_r\underline{x}_n)\}$ is still exchangeable in H_0. It has been shown that $\tilde{F}_n(y|G_n\underline{x}_n) \xrightarrow{p} F_n(y|G_n\underline{x}_n)$ as $n\rightarrow \infty$; see Romano (1989) and Pesarin (1989) for more details. Thus, if Conditions A and B of the above section are satisfied, the simulated version of $\phi_n^*(\underline{x}_n)$ has still the same asymptotic power as $\phi_n^c(\underline{x}_n)$.

If we assume that there a test with "best" power function exists in the class $\mathcal{C}$, this result gives a theoretical justification for the possible use of permutation techniques, i.e. conditional bootstrap methods, in the combination of dependent test statistics.

References.

BELL, C.B. and SEN, P.K. (1984). Randomization Procedures. In: *Handobook of Statistics* 4, 1-29. Krishnaiah, P.R., Sen, P.K. (eds.). Elsevier Science Publishers - North Holland, Amsterdam.

BERK, R.H. and JONES, D.H. (1978). Relatively Optimal Combinations of tests Statistics. *Scand. J. Statist.* 5, 159-162.

DWASS, M. (1955). On the Asymptotic Normality of Some Statistics Used in Non-Parametric Tests. *Ann. Math. Statist.* 26, 334-339.

DWASS, M. (1957). Modified Randomization Tests for Nonparametric Hypotheses. *Ann. Math. Statist.* 28, 181-187.

FOLKS, J.L. (1984). Combination of Independent Tests. In: *Handbook of Statistics* 4, 113-121. Krishnaiah, P.R., Sen, P.K. (eds). Elsevier science Publishers - North Holland, Amsterdam.

HINKLEY, D.V. (1989). Bootstrap Significance Tests. In: *Proc. 47th Session ISI - Paris* 53, 65-74.

HOEFFDING, W. (1952). The Large-Sample Power of tests Based on Permutations of Observations. *Ann. Math. Statist.* 23, 169-192.

LITTELL, R.C. and FOLKS, J.L. (1973). Asymptotic Optimality of Fisher's Method of Combining Independent Tests II. *J. Amer. Statist. Assoc.* 68, 193-194.

PESARIN, F. (1989). A Nonparametric Combination Method for Dependent Permutation Tests. Technical Report (1989:3), Dip.to di Sc. Statistiche, Università di Padova.

PURI, M.L. and SEN, P.K. (1971). *Nonparametric methods in multivariate Analysis.* Wiley & Sons, New York.

ROMANO, J.P. (1989). Bootstrap and Randomization Tests of some Nonparametric Hypotheses. *Ann. Statist.* 17, 141-159.

VADIVELOO, J. (1983). On the Theory of Modified Randomization test for Nonparametric Hypotheses. *Commun. Statist. - Theory Meth.* 12, 1581-1596.

Bootstrap for Time Series(5)

A Bootstrap Approach for Nonlinear Autoregressions

Some Preliminary Results

Jürgen Franke, University of Kaiserslautern
Matthias Wendel, Technical University of Berlin
Department of Mathematics, University of Kaiserslautern
P.O. Box 3049, D-6750 Kaiserslautern, F.R.G.

We consider *non-linear autoregressions of order 1* , i.e. discrete time processes generated by

$$X_{t+1} = m\,(X_t) + \epsilon_{t+1}, \quad -\infty < t < \infty, \tag{1}$$

where ϵ_t, $-\infty < t < \infty$, are i.i.d. zero-mean real random variables with probability density f_ϵ and finite variance $\sigma_\epsilon^2 = \mathrm{var}(\epsilon_t)$. Then, the transition probabilities $P(x, .)$ of the Markov chain (1) are absolutely continuous with density $f_\epsilon(.- m(x))$.

Further, we assume that $\{X_t, -\infty < t < \infty\}$ is *strictly stationary* and satisfies Rosenblatts *strong mixing* condition. The existence of such a solution of (1) follows from Doeblin's condition together with the requirement of aperiodicity and ergodicity (compare ch. VII of Rosenblatt, 1971). The integrability of ess $\sup_x$ $f_\epsilon(.- m(x))$ is sufficient for Doeblin's condition, and the other two conditions are valid and easily checked for a large class of regression functions m and laws of the ϵ_t. In the following, let f denote the marginal density of X_t.

Robinson (1983) has discussed nonparametric estimates for conditional expectations in a general framework. We consider the special case of estimating the regression function m from a finite sample $X_1, ..., X_n$ from (1). Here, the Nadaraya-Watson-type estimate, considered by Robinson, takes the form

$$\hat{m}(x;h) \;=\; \frac{1}{(n-1)h} \sum_{t=1}^{n-1} K\!\left[\frac{x-X_t}{h}\right] X_{t+1} \,/\, \hat{f}(x;h) \tag{2}$$

where $\hat{f}$ denotes a kernel estimate of f :

$$\hat{f}(x;h) \;=\; \frac{1}{nh} \sum_{t=1}^{n} K\!\left[\frac{x-X_t}{h}\right] . \tag{3}$$

We assume that K is a symmetric nonnegative kernel which integrates to 1 and whose tails are decreasing fast enough, i.e. for some $c, d, \beta > 0$:

$$0 \leq K(u) \leq c \,\exp(-\,d\,|u|^\beta), \quad -\infty < u < \infty .$$

If m and f are smooth enough, if the nonlinear autoregression $\{X_t, -\infty < t < \infty\}$ is mixing strongly enough, and if the bandwidth h converges to 0 for $n \longrightarrow \infty$ fast enough, then we know from Robinson (1983) that $\hat{m}(x;h)$ is a consistent asymptotically normal estimate of $m(x)$. More precisely, let us assume m is continuous on $(-\infty,\infty)$ and twice differentiable in x,   (4)

f is twice differentiable in x and $f(x) > 0$,

$E(X_t^4) < \infty$, and the mixing coefficients of the process (1) decrease at least as t^{-4}.

$nh \rightarrow \infty$, $nh^5 \rightarrow 0$ for $n \rightarrow \infty$.   (5)

Then, under the additional condition of local uniform (in s) boundedness of the bivariate densities of (X_t, X_{t+s}) , for $n \rightarrow \infty$:

$$(nh)^{\frac{1}{2}} \{\hat{m}(x;h) - m(x)\} \rightarrow Z \quad \text{in law,} \tag{6}$$

where Z is a Gaussian random variable with mean 0 and variance

$$\sigma_x^2 = \int_{-\infty}^{\infty} K^2(u)du \cdot \text{var}\{X_{t+1}|X_t=x\} / f(x)$$

$$= \int_{-\infty}^{\infty} K^2(u)du \cdot \sigma_\epsilon^2 / f(x)$$

Furthermore, σ_x^2 is consistently estimated by

$$\hat{\sigma}_x^2 = \int_{-\infty}^{\infty} K^2(u)du \cdot \hat{\sigma}_\epsilon^2 / \hat{f}(x;h) \quad \text{with} \quad \hat{\sigma}_\epsilon^2 = \frac{1}{n-1} \sum_{t=1}^{n-1} \{X_{t+1} - \hat{m}(X_t;h)\}^2 . \tag{7}$$

Further properties of the estimate $\hat{m}(x;h)$, e.g. uniform consistency, can be found in ch. III of Györfi et al. (1989). We remark that (6) and (7) remain valid if we replace $\hat{f}(x;h)$ in (2) and (7) by $\hat{f}(x;h')$ provided that the bandwidth sequence h' behaves asymptotically for $n \rightarrow \infty$ as h , in particular $\hat{f}(x;h)/\hat{f}(x;h')$ converges to 1 in probability. A good choice of bandwidth depends on the shape of m(x) and f(x) such that we can expect some improvement by allowing for different bandwidths in the numerator and denominator of (2).

The bootstrap procedure

We are interested in an approximation of the law of the estimate $\hat{m}(x;h)$ which improves upon asymptotic normality (6) for small and medium sample size. We propose the following bootstrap approximation which performs quite well in simulations. A theoretical foundation in the spirit of Härdle and Bowman (1988) and Franke and Härdle (1988) for the analogous problems of non-parametric regression and spectrum estimation will be the topic of a future paper. We start with an initial estimate $\hat{m}(x;g)$ of the form (2) where g denotes the bandwidth of the numerator and where we allow for choosing a different bandwidth g' for the denominator of (2). Then, we consider the empirical residuals and recenter them around 0:

$$\tilde{\epsilon}_{t+1} = X_{t+1} - \hat{m}(X_t;g) \quad , \quad \hat{\epsilon}_{t+1} = \tilde{\epsilon}_{t+1} - \frac{1}{n-1} \sum_{s=2}^{n} \tilde{\epsilon}_s \quad , \quad t = 1,...,\ n-1.$$

Then, we draw independent bootstrap residuals $\epsilon_1^*,...,\epsilon_n^*$ from the empirical distribution of $\hat{\epsilon}_2,...,\hat{\epsilon}_n$ and, finally, we get our bootstrap sample from

$$X_{t+1}^* = \hat{m}(X_t^*;g) + \epsilon_{t+1}^* \quad , \quad t = 0,...,n-1 . \tag{8}$$

There is the problem of stationarity connected with that of choosing the starting value X_o^*.

Essentially, there are two ways to overcome this difficulty. The first would be to use the information of the stationary distribution of X_o contained in the initial estimate $\hat{f}(x;g')$ of f(x).

We prefer to use an arbitrary X_o^* in the range covered by the sample $X_1,...,X_n$, then starting the recursion (8) but rejecting the first values until recourse to asymptotic stationarity gives us approximate stationarity.

Finally, we get the bootstrap estimate

$$\hat{m}^*(x;g,h) = \frac{1}{(n-1)h} \sum_{t=1}^{n-1} K\left[\frac{x-X_t^*}{h}\right] X_{t+1}^* \,/\, \hat{f}^*(x;g,h) \tag{9}$$

$$\hat{f}^*(x;g,h) = \frac{1}{nh} \sum_{t=1}^{n} K\left[\frac{x-X_t^*}{h}\right]$$

We use the conditional distribution of $(nh)^{\frac{1}{2}} \{\hat{m}^*(x;g,h) - \hat{m}(x;g)\}$ given the original data as an approximation for the distribution of $(nh)^{\frac{1}{2}} \{\hat{m}(x;h) - m(x)\}$.

Approximate confidence intervals

To judge the performance of the bootstrap approximation we consider the problem of finding a $(1-2\alpha)$ - confidence interval for $m(x)$, where x is fixed. Asymptotic normality (6) together with (7) implies the interval

$$\hat{m}(x;h) \pm (nh)^{-\frac{1}{2}} \hat{\sigma}_x z_{1-\alpha}, \tag{10}$$

where z_β, $0 < \beta < 1$, denotes the β-quantile of the standard normal law. As a bootstrap confidence interval, we consider the BC_a - interval of Efron (1987). We set

$$\zeta_\beta = \frac{b+z_\beta}{1-a(b+z_\beta)} \quad , 0 < \beta < 1 \quad ,$$

where as constants b and a we use

$$a = \frac{1}{6} \left(\sum_{t=2}^{n} \hat{\epsilon}_t^3 \right) / \left(\sum_{t=2}^{n} \hat{\epsilon}_t^2 \right)^{\frac{3}{2}} \quad , \quad \Phi(b) = G^*(\hat{m}(x;g)) \tag{11}$$

(compare Efron, 1987, ch. 7). Φ denotes the standard normal cdf, and G^* is the bootstrap cdf, i.e. the conditional cdf of $\hat{m}^*(x;g,h)$ given the original data. G^* and, therefore, b and ζ_β are not known, but we can approximate them by Monte Carlo sampling. In the following, think of b as such an approximation of the true bias-correcting constant, which we got from M Monte Carlo runs. Let $\hat{m}_i^*(x;g,h)$, $i = 1,...,M$, denote the bootstrap estimates from the Monte Carlo runs, and let $\hat{m}_{(\nu)}^*(x;g,h)$, $\hat{m}_{(\mu)}^*(x;g,h)$ be the two order statistics corresponding to $\nu/M \approx \Phi(b+\zeta_\alpha)$ and $\mu/M \approx \Phi(b+\zeta_{1-\alpha})$. Then, the bootstrap confidence interval

$$[\hat{m}_{(\nu)}^*(x;g,h) \,, \, \hat{m}_{(\mu)}^*(x;g,h)] \tag{12}$$

is a Monte Carlo-approximation of Efron's BC_a-interval

$$[(G^*)^{-1}(\Phi(b+\zeta_\alpha)) \,, \, (G^*)^{-1}(\Phi(b+\zeta_{1-\alpha}))]$$

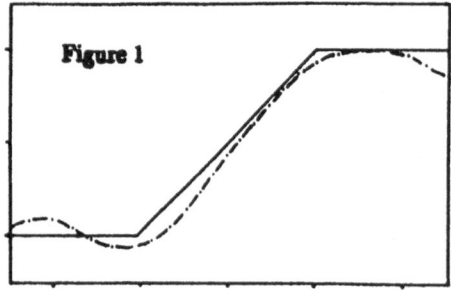

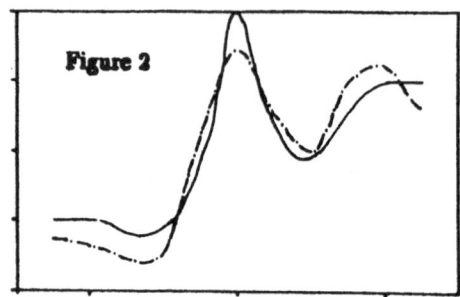

Simulation results

In our simulations, we have investigated several regression functions m, all of them coinciding with sgn(x) for $|x| \geq 1$, and several laws of the residuals ϵ_t. The sample size was n=200, and we chose the Gaussian kernel in (2) and (3). For calculating the BC_a-intervals (12) we used M=1000 bootstrap replications. We also considered "exact"$(1-2\alpha)$-confidence intervals which we got from a Monte Carlo simulation with 5000 runs. α was always 5%.

For the initial estimate $\hat{m}^*(x;g)$, we used different global (i.e. independent of x) bandwidths g for the numerator and g' for the denominator of (2), both chosen by a crossvalidatory argument (compare Härdle and Marron, 1985). As m is constant outside the unit interval, the global bandwidth g is rather large which seems to be advantageous for bootstrapping kernel estimates as observed by Franke and Härdle (1988). We calculated the "exact" confidence interval and the approximations (10) and (12) for $\hat{m}^*(x;h_x)$ for several values of x, where we allowed the bandwidths to depend on the location x. We made the following observations: compared to the "exact" interval the BC_a-interval (12) was quite satisfactory, compared to the asymptotic normal interval (10), which was considerably too large. As a conclusion, we give the details of two representative cases:

Case 1: $m(x) = x$ for $|x| \leq 1$; g = 0.367 , g' = 0.082 ; ϵ_t uniformly distributed with mean 0 and variance 1.

Case 2: $m(x)$ for $|x| \leq 1$ composed of cubic splines with constraints $m(\pm 1) = \pm 1$, m'(±1) = 0, m(0) = 2, m'(0) = 0, m'(1/8) = -10, m'(-1/8) = 10; g = 0.112, g' = 0.150; ϵ_t standard normally distributed.

Figure 1 and 2 show the regression function m(x) and the kernel estimate $\hat{m}^*(x;g)$ for both cases. Table 1 and 2 contain the values of x, which we considered, the bandwidths h_x and the "exact", asymptotic normal and BC_a-confidence intervals of level 90% for m(x). Remark that in these examples a was almost 0 such that the BC_a-intervals are close to the conventional bias-corrected bootstrap intervals. Further examples, showing the same general tendency, can be found in Wendel (1989).

Table 1

x	h_x	"exact"	confidence intervals: as. normal	BC_a
-2.5	0.224	-1.30,-0.55	-5.79, 3.96	-1.23,-0.63
-2.0	0.216	-1.24,-0.70	-4.09, 2.42	-1.03,-0.54
-1.5	0.217	-1.21,-0.74	-4.80, 2.66	-1.26,-0.83
-1.0	0.207	-1.07,-0.66	-4.35, 2.15	-1.28,-0.92
-0.5	0.297	-0.66,-0.27	-3.79, 2.29	-0.93,-0.60
0.0	0.209	-0.18, 0.18	-2.96, 2.69	-0.33, 0.03
0.5	0.298	0.28, 0.65	-2.70, 3.53	0.22, 0.60
1.0	0.200	0.65, 1.06	-2.76, 4.45	0.63, 1.01
1.5	0.222	0.74, 1.21	-2.88, 4.84	0.69, 1.19
2.0	0.222	0.70, 1.25	-4.37, 6.30	0.64, 1.23
2.5	0.234	0.56, 1.31	-7.01, 8.48	0.26, 1.14

Table 2

x	h_x	"exact"	confidence intervals: as. normal	BC_a
-1.25	0.067	-1.38,-0.61	-4.85, 2.30	-1.63,-0.84
-1.00	0.070	-1.37,-0.63	-4.45, 1.71	-1.73,-0.90
-0.75	0.069	-1.51,-0.80	-4.65, 1.56	-1.87,-1.12
-0.50	0.061	-1.44,-0.75	-5.03, 2.12	-1.80,-1.14
-0.25	0.065	-0.52, 0.19	-2.53, 2.89	-0.15, 0.59
0.00	0.056	1.32, 2.07	-1.06, 3.94	1.18, 1.78
0.25	0.066	0.10, 0.78	-2.04, 3.13	0.29, 0.90
0.50	0.067	-0.36, 0.31	-3.07, 3.07	-0.36, 0.29
0.75	0.065	0.15, 0.88	-2.00, 3.99	0.63, 1.31
1.00	0.069	0.57, 1.32	-1.84, 4.28	0.86, 1.58
1.25	0.063	0.58, 1.39	-2.65, 3.94	0.29, 0.97

References

Efron, B (1987), Better bootstrap confidence intervals (with discussion). JASA 82, 171-200.

Franke, J. and Härdle, W. (1988), On bootstrapping kernel spectral estimates. *Tentatively accepted for publication in Ann. Statist.

Györfi, L., Härdle, W., Sarda, P. and Vieu, Ph. (1989), Nonparametric Curve Estimation from Time Series. Lecture Notes in Statistics 60. Springer, Berlin-Heidelberg-New York.

Härdle, W. and Bowman, A. (1988), Bootstrapping in nonparametric regression: Local adaptive smoothing and confidence bands. JASA 83, 102-110.

Härdle, W. and Marron, J.S. (1985), Optimal Bandwidth Selection in Nonparametric Regression Function Estimation. Ann. Statist., 13, 1465-1481.

Robinson, P.M. (1983), Nonparametric estimators for time series. J. Time Series Anal. 4, 185-207.

Rosenblatt, M. (1971), Markov Processes. Structure and Asymptotic Behavior. Springer, Heidelberg.

Wendel, M. (1989), Eine Anwendung der Bootstrap - Methode auf nicht-lineare autoregressive Prozesse erster Ordnung. Diploma thesis, Technical University of Berlin.

Bootstrap procedures for AR (∞) - processes

Jens-Peter Kreiss

Technical University Braunschweig

In this paper we will deal with an application of Efron's 1979 bootstrap to stationary stochastic processes in discrete time. In many applications it is assumed that these processes are of autoregressive or more generally of autoregressive moving average type, i.e. the underlying stationary process $\mathbf{X} = (X_t : t \in \mathbf{Z} = \{0, \pm 1, \pm 2, ...\})$ is assumed to satisfy the following stochastic difference equation

$$X_t = \sum_{\nu=1}^{p} a_\nu X_{t-\nu} + \varepsilon_t + \sum_{\mu=1}^{q} b_\mu \varepsilon_{t-\mu} \, , \ t \in \mathbf{Z}.$$

Here $\varepsilon = (\varepsilon_t : t \in \mathbf{Z})$ denotes a white noise, that is a sequence of uncorrelated, zero mean random variables with finite variance σ^2.

One relevant statistical issue is the estimation of the parameter $\vartheta = (\sigma^2, a_1, ..., a_p, b_1, ..., b_q)$. For the construction of estimators $\hat{\vartheta}_n$ it is often assumed that the order p and q of the model is known.

As can be found in papers of Efron and Tibshirani (1986) or Bose (1988) for purely autoregressive models (i.e. $q = 0$) and Franke and Kreiss (1989) for the general case, the bootstrap principle yields a consistent estimate of the distribution $\mathcal{L}(\sqrt{n}(\hat{\vartheta}_n - \vartheta))$. One necessary assumption for such a result is an i.i.d.-structure of the white noise ε.

In applications it is now rarely the case that the order of the underlying model is known. Moreover, autoregressive or autoregressive moving average models should be understood as an approximation to the true underlying situation, only. One proposal, which we intend to discuss in detail, is to assume that the *true* model belongs to a reasonable large class of stationary processes,which contains at least all autoregressive moving average models. The autoregressive processes with infinite order $[AR(\infty)]$ may serve as such a large class of stationary processes. This means $\mathbf{X}$ fulfills

$$X_t = \sum_{\nu=1}^{\infty} a_\nu X_{t-\nu} + \varepsilon_t \, , \ t \in \mathbf{Z}.$$

As is indicated above we will assume that ε_t , $t \in \mathbf{Z}$, are i.i.d. with zero mean and finite variance $\sigma^2 > 0$. Further assume that $1 - \sum_{\nu=1}^{\infty} a_\nu z^\nu$ is holomorphic in an open neighborhood of the closed unit disk and has no zeroes with magnitude less than or equal to one.

For the sake of simplicity assume that our interest lies in the estimation of the following one-dimensional functional of the parameter $\vartheta = (\sigma^2, a_\nu : \nu \in \mathbb{N})$

$$T(\vartheta) = a_1.$$

For example, an estimator for a_1 is wanted, if we intend to fit a first order moving average model to the given set of data. The optimal moving average coefficient (which we have to estimate) is just a_1 (cf. Kreiss (1988), Example 2.2).

On the basis of observations $X_1, ..., X_n$ we intend to construct an estimator for $T(\vartheta)$. To do so, we fit in a first place an autoregressive scheme of order p, where $p = p(n)$ should depend at least on the sample size n. This can be done for example by using the usual Yule-Walker estimator

$$(\hat{a}_{1,n}(p), ..., \hat{a}_{p,n}(p))^T = \hat{\Gamma}_n(p)^{-1} \frac{1}{n-p} \sum_{j=p+1}^{n} X_j \begin{pmatrix} X_{j-1} \\ \vdots \\ X_{j-p} \end{pmatrix},$$

where

$$\hat{\Gamma}_n(p) = \left(\frac{1}{n-p} \sum_{j=p+1}^{n} X_{j-r} X_{j-s} \right)_{r,s=1,...,p}$$

Our main interest is to construct a bootstrap-approximation of the following distribution

$$\mathcal{L}\left(\sqrt{n}(\hat{a}_{1,n}(p) - a_1) \right) =: \mathcal{L}_n.$$

Before doing so, let us have a brief look on the asymptotic distribution of this expression for a very special situation.

Lemma: Assume $a_\nu = -\rho^\nu, -1 < \rho < 1, p \to \infty$ and $p^3/n \to 0$ as $n \to \infty$. If $\sqrt{n}\rho^{2p} \to C$ as $n \to \infty$ then

$$\mathcal{L}\left(\sqrt{n}(\hat{a}_{1,n}(p) - a_1) \right) \Rightarrow \mathcal{N}(C \cdot \rho(1 - \rho^2), 1).$$

Proof: With $u = (1, 0, ..., 0)^T \in \mathbb{R}^p$ and $\Gamma(p) = (EX_s X_t : s, t = 1, ..., p) = (\gamma(s - t) : s, t = 1, ..., p)$ we obtain

$$\sqrt{n}(\hat{a}_{1,n}(p) - a_1)$$

$$= \sqrt{n}\left(u^T \hat{\Gamma}_n(p)^{-1} \frac{1}{n-p} \sum_{j=p+1}^{n} X_j (X_{j-1}, ..., X_{j-p})^T - a_1 \right)$$

$$= \sqrt{n}\, u^T \left(\hat{\Gamma}_n(p)^{-1} \frac{1}{n-p} \sum_{j=p+1}^{n} X_j (X_{j-1}, ..., X_{j-p})^T - (a_1, ..., a_p)^T \right)$$

$$= \sqrt{\frac{n}{n-p}}\, u^T \hat{\Gamma}_n(p)^{-1} \frac{1}{\sqrt{n-p}} \sum_{j=p+1}^{n} \varepsilon_j (X_{j-1}, ..., X_{j-p})^T$$

$$+ \; u^T \hat{\Gamma}_n(p)^{-1} \frac{\sqrt{n}}{n-p} \sum_{j=p+1}^{n} \sum_{\nu=p+1}^{\infty} a_\nu X_{j-\nu} (X_{j-1}, ..., X_{j-p})^T.$$

Use Lemma 1 in Kreiss (1988), p. 87, the explicit expression for $\Gamma(p)^{-1}$ in Lemma 2, Kreiss (1988), p. 88, and a martingale central limit theorem, cf. Brown (1971), to obtain for the first summand

$$u^T \hat{\Gamma}_n(p)^{-1} \frac{1}{\sqrt{n-p}} \sum_{j=p+1}^{n} \varepsilon_j (X_{j-1}, ..., X_{j-p})^T$$

$$= \frac{1}{\sqrt{n-p}} \sum_{j=p+1}^{n} \varepsilon_j u^T \Gamma(p)^{-1} (X_{j-1}, ..., X_{j-p})^T + o_P(1) \Rightarrow \mathcal{N}(0,1).$$

Use the above mentioned results of Kreiss again to obtain for the second summand

$$u^T \hat{\Gamma}_n(p)^{-1} \frac{\sqrt{n}}{n-p} \sum_{j=p+1}^{n} \sum_{\nu=p+1}^{\infty} a_\nu X_{j-\nu} (X_{j-1}, ..., X_{j-p})^T$$

$$= \frac{\sqrt{n}}{n-p} u^T \Gamma(p)^{-1} \sum_{j=p+1}^{n} \sum_{\nu=p+1}^{\infty} a_\nu X_{j-\nu} (X_{j-1}, ..., X_{j-p})^T + o_P(1)$$

$$= \sqrt{n}\, u^T \Gamma(p)^{-1} \sum_{\nu=p+1}^{\infty} a_\nu (\gamma(1-\nu), ..., \gamma(p-\nu))^T + o_P(1).$$

From $X_t = \sum_{\nu=1}^{\infty} a_\nu X_{t-\nu} + \varepsilon_t$ we obtain for $k = 1, ..., p$ the equality $\gamma(k) - \sum_{\nu=1}^{p} a_\nu \gamma(k-\nu) = \sum_{\nu=p+1}^{\infty} a_\nu \gamma(k-\nu)$. This implies that the above expression equals

$$\sqrt{n}\, u^T \Gamma(p)^{-1} \Big(\gamma(k) - \sum_{\nu=1}^{p} a_\nu \gamma(k-\nu) \Big)_{k=1,...,p} + o_P(1)$$

$$= \sqrt{n} \Big(u^T \Gamma(p)^{-1} (\gamma(1), ..., \gamma(p))^T - a_1 \Big) + o_P(1),$$

since $\Gamma(p)^{-1} = (\gamma(r-s) : r, s = 1, ..., p)^{-1}$. Assume that $(a_1(p), ..., a_p(p))$ is the value which minimizes $E(X_p - \sum_{\nu=1}^{p} c_\nu X_{p-\nu})^2$ to obtain, that the last expression is equal to $\sqrt{n}(a_1(p) - a_1)$, which describes the bias of the distribution under consideration.

For the special case $a_\nu = -\rho^\nu$, $-1 < \rho < 1$, we are able to compute the bias, namely

$$\sqrt{n}(a_1(p) - a_1) = \sqrt{n} \Big(u^T \Gamma(p)^{-1} (\gamma(1), ..., \gamma(p))^T + \rho \Big) = \sqrt{n} \frac{\rho^{2p+1}}{\sum\limits_{\nu=0}^{p} \rho^{2\nu}}.$$

From this we obtain the desired result. $\square$

Now we will deal with the bootstrap procedure in order to approximate the distribution $\mathcal{L}_n$. We propose the following **bootstrap procedure.**

1. Step: Use observations $X_1, ..., X_n$ of the AR (∞) - process to fit an autoregressive scheme of order $p_{Bootstrap}(n) = p_o$. Using again the Yule-Walker procedure we obtain

$$(\hat{a}_{1,n}(p_0), ..., \hat{a}_{p_0,n}(p_0)).$$

2. Step: Compute estimated residuals

$$\hat{\varepsilon}_{t,n} := X_t - \sum_{\nu=1}^{p_0} \hat{a}_{\nu,n}(p_0) X_{t-\nu}, \quad t = p_0 + 1, ..., n,$$

and consider the corresponding *centered* empirical cdf

$$\hat{F}_n(x) = \frac{1}{n-p_0} \sum_{t=p_0+1}^{n} 1_{(\hat{\varepsilon}_{t,n} - \hat{\varepsilon}_{\bullet,n} \leq x)}, \quad x \in \mathbb{R}, \quad \hat{\varepsilon}_{\bullet,n} = \frac{1}{n-p_0} \sum_{t=p_0+1}^{n} \hat{\varepsilon}_{t,n}.$$

3. Step: Create i.i.d. random variables

$$e^*_{-Q}, ..., e^*_n , \ Q \in \mathbb{N},$$

distributed according to $\hat{F}_n$. This is the usual bootstrap sample of residuals.

4. Step: Compute according to

$$X^*_{t,n} = \sum_{\nu=1}^{p_0} \hat{a}_{\nu,n}(p_0) X^*_{t-\nu,n} + e^*_t , \ t = -Q, ..., n,$$

where the initial values $X^*_{-1-Q,n} = \cdots = X^*_{-p_0-Q,n} = 0$ are used, a bootstrap sample of the time series itself, and obtain the **bootstrap estimator**

$$(\hat{a}^*_{1,n}(p), ..., \hat{a}^*_{p,n}(p)).$$

The bootstrap hypothesis now is, that $\mathcal{L}_n$ is reasonably well approximated (at least asymptotically) by

$$\mathcal{L}_n(X_1, ..., X_n) := \mathcal{L}\left(\sqrt{n}(\hat{a}^*_{1,n}(p) - \hat{a}_{1,n}(p_0)) \middle| X_1, ..., X_n\right).$$

We will prove the asymptotic validity of the bootstrap. Asymptotic validity in this context means, that the conditional distribution $\mathcal{L}_n(X_1, ..., X_n)$ converges in distribution to the same limit law as $\mathcal{L}_n$ does (in probability). In extension to Kreiss (1988), section four, we are in this paper interested, whether or not the bootstrap procedure reflects the asymptotical bias occuring for example in the special situation of the lemma, if $\sqrt{n}\rho^{2p} \rightarrow_{n\rightarrow\infty} C \neq 0$.

But before we give a formal proof of such a result, let us consider a small simulation study. This study demonstrates the quite well behavior of the bootstrap even in finite samples, at least for the special model of the lemma.

Example: Consider the same simple AR(∞)-model as in the lemma, i.e.

$$X_t = -\sum_{\nu=1}^{\infty} \rho^\nu X_{t-\nu} + \varepsilon_t , \ t \in \mathbb{Z} , \ \rho = 0.80 , \ \varepsilon_1 \sim \mathcal{N}(0,1).$$

Of course $(X_t : t \in \mathbb{Z})$ is nothing else than a first order moving average process with coefficient $-\rho$. We simulated the distribution of $\sqrt{n}(\hat{a}_{1,n}(p) + \rho)$ for $n = 100$ observations and fitted autoregressive model of order $p = 3$. Notice, that we make no use of the moving average structure of the data. In *figure 1* we plotted a histogram of this distribution (solid line) obtained by 3000 Monte Carlo runs.

The dotted and dashed line in figure 1 corresponds to two simulated bootstrap distributions $\mathcal{L}_n(X_1, ..., X_n) = \mathcal{L}\left(\sqrt{n}(\hat{a}^*_{1,n}(p) - \hat{a}_{1,n}(p_o)) \middle| X_1, ..., X_n\right)$ [again $n = 100, p = 3$ and 3000 Monte Carlo replications]. To perform the bootstrap procedure we fitted an autoregressive model of order $p_o = 8$ to the original data.

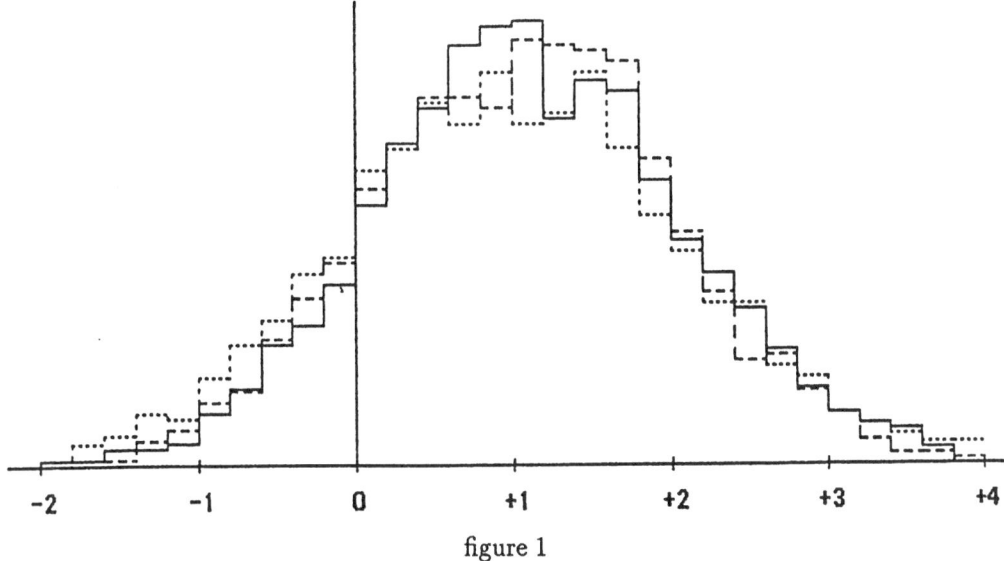

figure 1

Finally we formulate and prove the asymptotic validity of the bootstrap for general $AR(\infty)$-processes.

Theorem: *Assume the $AR(\infty)$-model with i.i.d. random variables $(\varepsilon_t : t \in \mathbf{Z})$, $E\varepsilon_1 = 0$, $E\varepsilon_1^2 = \sigma^2 \in (0, \infty)$ and $E\varepsilon_1^4 < \infty$. Further assume that $\rho \in (0, 1)$ exists, such that the co-efficients of the following two holomorphic functions $1 - \sum_{\nu=1}^{\infty} a_\nu z^\nu$ and $(1 - \sum_{\nu=1}^{\infty} a_\nu z^\nu)^{-1} = 1 + \sum_{\nu=1}^{\infty} \alpha_\nu z^\nu$ fullfill for suitable constants $C_1, C_2 > 0$:*

$$|a_\nu| \leq C_1 \cdot \rho^\nu \ , \ |\alpha_\nu| \leq C_2 \cdot \rho^\nu \text{ for all } \nu \in \mathbf{N}.$$

$(p_o = p_o(n) : n \in \mathbf{N})$ and $(p = p(n) : n \in \mathbf{N})$ denote two sequences of positive integers converging to infinity with $p_o \sim n^\alpha (\alpha > 0)$, $p_o^4/n \to 0$, $p/n^\delta \to 0$ (for all $\delta > 0$) and $p\sqrt{n}\rho^{2p}$ bounded as $n \to \infty$. Under these assumptions we obtain the asymptotic validity of the bootstrap.

For the special situation of the lemma and the example we get the asymptotic validity of the bootstrap under the weaker assumption $\sqrt{n}\rho^{2p} \to C$ as $n \to \infty$ instead of $p\sqrt{n}\rho^{2p}$ bounded, that is

$$\mathcal{L}\left(\sqrt{n}(\hat{a}_{1,n}^*(p) - \hat{a}_{1,n}(p_o)) \middle| X_1, ..., X_n\right) \Rightarrow \mathcal{N}(C \cdot \rho(1 - \rho^2), 1) \text{ in probability}.$$

This means that the bootstrap reflects the asymptotical bias for this very special situation.

Proof: (We give an outline of the proof, only). For $u^T = (1, 0, ..., 0) \in \mathbb{R}^p$ consider

$$\sqrt{n}(\hat{a}_{1,n}^*(p) - \hat{a}_{1,n}(p_o))$$

$$= u^T \hat{\Gamma}_n^*(p)^{-1} \frac{\sqrt{n}}{n-p} \sum_{j=p+1}^{n} \left(X_{j,n}^* - \sum_{\nu=1}^{p} \hat{a}_{\nu,n}(p_o) X_{j-\nu,n}^*\right)(X_{j-1,n}^*, ..., X_{j-p,n}^*)^T.$$

Since $\|(\hat{a}_{1,n}(p_o) - a_1, ..., \hat{a}_{p_o,n}(p_o) - a_{p_o})\|^2 = \mathcal{O}_P(\frac{p_o}{n})$ and for $(1 - \sum_{\nu=1}^{p_o} \hat{a}_{\nu,n}(p_o)z^\nu)^{-1} = 1 + \sum_{\nu=1}^{\infty} \hat{\alpha}_{\nu,n}(p_o)z^\nu$:

$$\sup_{\nu \in \mathbf{N}}(1 + \frac{1}{p_o})^\nu |\hat{\alpha}_{\nu,n}(p_o) - \alpha_\nu| = \mathcal{O}_P(\frac{p_o}{\sqrt{n}}),$$

cf. Kreiss (1988) [Lemma 1.4 and Lemma 4 of Appendix three], we obtain [abbreviate $E^*(\cdot) = E(\cdot|X_1, ..., X_n)$] for suitable $\delta > 0$

$$E^*\left\|\hat{\Gamma}_n^*(p) - \left(\sum_{\mu=0}^{\infty} \alpha_\mu \alpha_{\mu+r-s} \cdot E^*(e_1^*)^2 \ : \ r, s = 1, ..., p\right)\right\|^2 = o_P(n^{-1/2-\delta}).$$

This implies that the asymptotic behaviour of $\sqrt{n}(\hat{a}_{1,n}^*(p) - \hat{a}_{1,n}(p_o))$ is equivalent to that of

$$\frac{1}{E^*(e_1^*)^2} \cdot \frac{1}{\sqrt{n-p}} \sum_{j=p+1}^{n} \left(X_{j,n}^* - \sum_{\nu=1}^{p} \hat{a}_{\nu,n}(p_o) X_{j-\nu,n}^*\right) \sum_{r=1}^{p} \kappa_r(p) X_{j-r,n}^*,$$

where $u^T\left(\sum_{\mu=0}^{\infty} \alpha_\mu \alpha_{\mu+r-s} \ : \ r, s = 1, ..., p\right)^{-1} = (\kappa_1(p), ..., \kappa_p(p))$. Since $E^*(e_1^*)^2 \to \sigma^2$ in probability, cf. Kreiss (1988) [Corollary 3.6], it suffices to consider the second factor, which equals (without loss of generality assume $Q = 0$ in **Step 3** of our bootstrap proposal)

$$\frac{1}{\sqrt{n-p}} \sum_{j=p+1}^{n} \left(e_j^* + \sum_{\nu=p+1}^{j} (\hat{\alpha}_{\nu,n}(p_o) - \sum_{\mu=1}^{p} \hat{a}_{\mu,n}(p_o)\hat{\alpha}_{\nu-\mu,n}(p_o))e_{j-\nu}^*\right) \sum_{r=1}^{p} \kappa_r(p) X_{j-r,n}^*$$

$$= \frac{1}{\sqrt{n-p}} \sum_{j=p+1}^{n} \left(e_j^* + \sum_{\nu=p+1}^{j} (\sum_{\mu=p+1}^{p_o} \hat{a}_{\mu,n}(p_o)\hat{\alpha}_{\nu-\mu,n}(p_o))e_{j-\nu}^*\right) \sum_{r=1}^{p} \kappa_r(p) X_{j-r,n}^*.$$

Using our assumptions and $|\kappa_r(p)| \le C_o \cdot (\rho^\nu + \rho^p)$, cf. Kreiss (1988) [Corollary 3 of Appendix three], we obtain after long and tedious but direct computation, that the expectation $E^*(\cdot)$ of the absolute value of the difference of this expression with the following one converges to zero in probability

$$U_n(X_1, ..., X_n) := \frac{1}{\sqrt{n-p}} \sum_{j=p+1}^{n} \left(e_j^* + \sum_{\nu=p+1}^{j} (\sum_{\mu=p+1}^{p_o} a_\mu \alpha_{\nu-\mu})e_{j-\nu}^*\right) \sum_{r=1}^{p} \kappa_r(p) \sum_{\tau=0}^{j-r} \alpha_\tau e_{j-r-\tau}^*.$$

Consider for example

$$E^*\left|\frac{1}{\sqrt{n-p}} \sum_{j=p+1}^{n} \sum_{\nu=p+1}^{j} (\sum_{\mu=p+1}^{p_o} (\hat{a}_{\mu,n}(p_o) - a_\mu)\hat{\alpha}_{\nu-\mu,n}(p_o))e_{j-\nu}^* \sum_{r=1}^{p} \kappa_r(p) X_{j-r,n}^*\right|$$

$$\le \sum_{\mu=p+1}^{p_o} |\hat{a}_{\mu,n}(p_o) - a_\mu| \sum_{r=1}^{p} |\kappa_r(p)| \left(E^*\left(\frac{1}{\sqrt{n-p}} \sum_{j=p+1}^{n} \sum_{\nu=p+1}^{j} \hat{\alpha}_{\nu-\mu,n}(p_o)e_{j-\nu}^* X_{j-r,n}^*\right)^2\right)^{1/2}$$

$$\le \sqrt{p_o} \, \|(\hat{a}_{1,n}(p_o) - a_1, ..., \hat{a}_{p_o,n}(p_o) - a_{p_o})\| \cdot \mathcal{O}_P\left(\frac{p_o^3}{\sqrt{n}} + p\rho^p \sqrt{n}\right) = o_P(1),$$

and

$$E^*\left|\frac{1}{\sqrt{n-p}} \sum_{j=p+1}^{n} \sum_{\nu=p+1}^{j} \sum_{\mu=p+1}^{p_o} a_\mu(\hat{\alpha}_{\nu-\mu,n}(p_o) - \alpha_{\nu-\mu})e_{j-\nu}^* \sum_{r=1}^{p} \kappa_r(p) X_{j-r,n}^*\right|$$

$$\le \sum_{\mu=p+1}^{p_o} |a_\mu| \sum_{r=1}^{p} |\kappa_r(p)| \left(E^*\left(\frac{1}{\sqrt{n-p}} \sum_{j=p+1}^{n} \sum_{\nu=p+1}^{j} (\hat{\alpha}_{\nu-\mu,n}(p_o) - \alpha_{\nu-\mu})e_{j-\nu}^* X_{j-r,n}^*\right)^2\right)^{1/2}$$

$$\le \mathcal{O}_P(1) \cdot \sqrt{\rho^{2p}\sqrt{n}} \left(\frac{p_o^4}{n}\right)^{3/4} = o_P(1).$$

Along the same lines we obtain that the following expression has the same asymptotic distribution as $\sqrt{n}(\hat{a}_{1,n}(p) - a_1)$

$$V_n := \sigma^{-2} \cdot \frac{1}{\sqrt{n-p}} \sum_{j=p+1}^{n} \Big(\varepsilon_j + \sum_{\nu=p+1}^{j} \big(\sum_{\mu=p+1}^{p_0} a_\mu \alpha_{\nu-\mu} \big) \varepsilon_{j-\nu} \Big) \sum_{r=1}^{p} \kappa_r(p) \sum_{\tau=0}^{j-r} \alpha_\tau \varepsilon_{j-r-\tau} \ .$$

Because of this we can restrict our further considerations to these two terms. Using Mallow's metric (cf. Bickel and Freedman (1981)), i.e.

$$d_2(P,Q) := \inf \big(E|X - Y|^2 \big)^{1/2} \ ,$$

where the infimum covers all pairs of real valued random variables (X,Y) which have marginal distributions P, Q, respectively, we eventually obtain

$$\frac{1}{2} \cdot d_2 \big(U_n(X_1, ..., X_n), V_n \big)^2$$

$$\leq \ \inf \Big[E \Big(\frac{1}{\sqrt{n-p}} \sum_{j=p+1}^{n} \big(e_j^* - \varepsilon_j + \sum_{\nu=p+1}^{j} \sum_{\mu=p+1}^{p_0} a_\mu \alpha_{\nu-\mu}(e_{j-\nu}^* - \varepsilon_{j-\nu}) \big) \sum_{r=1}^{p} \kappa_r(p) \sum_{\tau=0}^{j-r} \alpha_\tau e_{j-r-\tau}^* \Big)^2$$

$$+ \ E \Big(\frac{1}{\sqrt{n-p}} \sum_{j=p+1}^{n} \big(\varepsilon_j + \sum_{\nu=p+1}^{j} \big(\sum_{\mu=p+1}^{p_0} a_\mu \alpha_{\nu-\mu} \big) \varepsilon_{j-\nu} \big) \sum_{r=1}^{p} \kappa_r(p) \sum_{\tau=0}^{j-r} \alpha_\tau (e_{j-r-\tau}^* - \varepsilon_{j-r-\tau}) \big)^2 \Big] \ ,$$

where the infimum covers all sequences $(e_1^*, \varepsilon_1), ..., (e_n^*, \varepsilon_n)$ of i.i.d. pairs of random variables with marginal distributions $\hat{F}_n, F$, respectively. F denotes the distribution function of ε_1. Both expressions can be bounded through

$$n \cdot \rho^{2p} \cdot E \, (e_1^* - \varepsilon_1)^2 \ .$$

Since $\inf E \, (e_1^* - \varepsilon_1)^2 = d_2(\hat{F}_n, F)^2 \to 0$ in probability, cf. Kreiss (1988) [Corollary 3.6], we obtain from Bickel and Freedman (1981) [Lemma 8.3] the desired result for the general case. For the special situation we obtain instead of the last bound $n \cdot \rho^{4p} \cdot E \, (e_1^* - \varepsilon_1)^2$. □

References

Bickel, P.J. and Freedman, D.A. (1981). Some Asymptotic theory for the bootstrap. *Ann. Statist.* **9**, 1196-1217.

Bose, A. (1988). Edgeworth corrections by bootstrap in autoregressions. *Ann. Statist.* **16**, 1709-1722.

Brown, B.M. (1971). Martingale central limit theorems. *Ann. Math. Statist.* **42**, 54-66.

Efron, B. and Tibshirani, R. (1986). Bootstrap methods for standard errors, confidence intervals, and other measures of statistical accuracy. *Statist. Science* **1**, 54-77.

Franke, J. and Kreiss, J.-P. (1989). Bootstrapping stationary ARMA models. *Preprint.* Submitted.

Freedman, D.A. (1984). On bootstrapping two-stage least-squares estimates in stationary linear models. *Ann. Statist.* **12**, 827-842.

Kreiss, J.-P. (1988). Asymptotic statistical inference for a class of stochastic processes. *Habilitationsschrift.* University of Hamburg.

BOOTSTRAPPING SOME STATISTICS USEFUL IN IDENTIFYING ARMA MODELS

Efstathios Paparoditis

Fachbereich Wirtschaftswissenschaften, Freie Universität Berlin

Garystrasse 21 , D-1000 Berlin 33

1 Introduction

Consider a zero mean weakly stationary stochastic process with a continuous and nonzero spectral density function, which satisfies the stochastic difference equation

$$X_t = \sum_{j=1}^{\infty} a_j X_{t-j} + \varepsilon_t, \tag{1}$$

for $t \epsilon \mathbf{Z}$. We assume that the associated power series $A(z) = 1 - \sum_{j=1}^{\infty} a_j z^j$ converges and is nonzero for $|z| \leq 1$. The random variables ε_t are assumed to be independently and identically distributed according to an unknown distribution function F with $E\varepsilon_t = 0$ and $E\varepsilon_t^2 = \sigma^2 > 0$.

We are interested in the estimation of the distribution of some statistics which are useful in identifying the orders p and q, when (1) has an ARMA(p,q) representation. Denote by $\Gamma(i,j)$ the $i \times i$ Toeplitz matrix $\Gamma(i,j) = (\gamma_{j-m+n})_{m,n=1}^{i}$, where $\gamma_h = E(X_t X_{t+h})$ is the autocovariance at lag $h \epsilon \mathbf{Z}$ of the process considered. Let $\lambda(i,j)$ be the vector autocorrelation of order $i \epsilon \mathbf{N}$ and lag $j \epsilon \mathbf{Z}$ of the process defined by

$$\lambda(i,j) = (-1)^{i-1} \frac{det\Gamma(i,j)}{det\Gamma(i,0)}. \tag{2}$$

Notice that $\lambda(1,j)$ is the well-known autocorelation of lag j and $\lambda(i,1)$ is the partial autocorrelation of lag i of the process considered. If we arrange the vector autocorrelations $\lambda(i,j)$ in a two dimensional table with rows indexed by $i \epsilon \mathbf{N}$ and columns indexed by $j \epsilon \mathbf{N}$ then the following identification property of vector autocorrelations is true: $(X_t; t \epsilon \mathbf{Z})$ is an ARMA(p,q) process, iff $\lambda(i,j) \neq 0$ for $i = p$, $j \geq q$ and $j = q$, $i \geq p$; $\lambda(i,j) = 0$ for all $i \geq p+1$ and $j \geq q+1$. For a proof of this order identification property see Streitberg(1982). For more details about vector autocorrelations and their connections to other order identifications approaches see Paparoditis(1990).

Now, let $X_1, X_2, \ldots, X_N$ be a realisation of the process considered and denote by $\hat{\Gamma}(i,j)$ the same matrix as $\Gamma(i,j)$ obtained after replacing γ_h by the sample autocovariance $\hat{\gamma}_h = N^{-1} \sum_{t=1}^{N-h} X_t X_{t+h}$. It can be shown, that if $E(\varepsilon^4) < \infty$ the distribution of $\sqrt{N}(\hat{\lambda}(i,j) - \lambda(i,j))$ converges to a normal distribution with mean zero and variance $\tau^2(i,j)$, cf. Paparoditis(1990). However, since the considered statistics are strongly nonlinear, we expect that their finite sample distribution is not well approximated by the normal.

2 A parametric-bootstrap algorithm

Since we have time dependent data we cannot choose single values with replacement as in the i.i.d. case. Thus we have to develop a resample algorithm which is able to reproduce the dependence structure of the observed process even if the true model class is unknown. It is well known that for every nonsingular covariance matrix $\Gamma(k+1,0)$ there is an AR(k) process whose autocovariances at lag $0,1,\ldots,k$ are $\gamma_0,\gamma_1,\ldots,\gamma_k$. For instance if we fit an autoregressive process of order m to the data by means of the Yule-Walker estimates, then the autocorrelations of this processes up to lag m equal the sample autocorrelations of the observed process. Thus we can use an autoregressive process of high-order in order to reproduce the dependence structure of the data and to generate the bootstrap samples. In particular the proposed Bootstrap procedure generates samples $(X_1^*, X_2^*, \ldots, X_N^*)$ of the process (1) through the following steps.

First we find estimates $\hat{\varepsilon}$ of the innovations by fitting a high-order autoregression to the data, i.e. we approximate the process considered by an AR(m), $m < \infty$ process. Let $m = m(N) \to \infty$ for $N \to \infty$ and denote by $\hat{a}_N(m) = (\hat{a}_{1,N}, \hat{a}_{2,N}, \ldots, \hat{a}_{m,N})^T$ an arbitrary sequence of estimators of the first m coefficients $a(m) = (a_1, a_2, \ldots, a_m)^T$, which satisfies the condition

$$\|\hat{a}_N(m) - a(m)\|^2 = O_p(\frac{m}{N}). \tag{3}$$

Such estimates may be in practice the Yule-Walker estimates or the least squares estimates. From this we obtain estimated residuals

$$\hat{\varepsilon}_t = X_t - \sum_{i=1}^{m} \hat{a}_{i,N} X_{t-i}$$

for $t = m+1, \ldots, N$. Now, let $\hat{F}_N$ be the empirical distribution of $\hat{\varepsilon}_t - \hat{\varepsilon}.$, where $\hat{\varepsilon}. = (N - m)^{-1} \sum_{t=m+1}^{N} \hat{\varepsilon}_t$. Denote further by $\{\hat{b}_{j,N}; j\epsilon\mathbf{N}\}$ the coefficients of the MA(∞) representation of the fitted AR(m) process, i.e. $1 + \sum_{j=1}^{\infty} \hat{b}_{j,N} z^j = (1 - \sum_{i=1}^{m} \hat{a}_{i,N} z^i)^{-1}$, $\mid z \mid \leq 1$ and let $\delta > 0$ be a desired error level. Select $s\epsilon\mathbf{N}$ sufficiently large such that $\gamma_{0,m} - \sum_{j=0}^{s} \hat{b}_{j,N}^2 < \delta$. The quantity $\gamma_{0,m}$ denotes the theoretical variance of the fitted AR(m) process, when the error variance equals one.

Now, generate a sample of length $N+s+1$ of random variables e_j, where the e_j are independently and identically distributed according to $\hat{F}_N$

$$e_{-s}, e_{-s+1}, e_{-s+2}, \ldots, e_0, e_1, e_2, \ldots, e_n.$$

The first m values are calculated according to

$$X_t^* = \sum_{j=0}^{s+1} \hat{b}_{j,N} e_{t-j},$$

for $t = 1, \ldots, m$, where $b_{0,N} = 1$. The remaining values for $t = m+1, m+2, \ldots, N$ are then determined from the equation

$$X_t^* = \sum_{j=1}^{m} \hat{a}_{j,N} X_{t-j}^* + e_t.$$

Denote by $\hat{\lambda}^*(i,j)$ the same statistic as (2) calculated using the bootstrap sample $(X_1^*, X_2^*, \ldots, X_N^*)$ and let $\lambda^*(i,j) = E[\hat{\lambda}^*(i,j)]$. The finite sample distribution of the statistic $\sqrt{N}(\hat{\lambda}(i,j) - \lambda(i,j))$ can be approximated by the distribution of the statistic $\sqrt{N}(\hat{\lambda}^*(i,j) - \lambda^*(i,j))$. The distribution of this statistic has to be evaluated by simulation. The asymptotic validity of this approximation is established by the following theorem, which shows that the conditional distribution of $\sqrt{N}(\hat{\lambda}^*(i,j) - \lambda^*(i,j))$ given the observed part of the process considered, and the unconditional distribution of $\sqrt{N}(\hat{\lambda}(i,j) - \lambda(i,j))$ have the same limit.

Theorem 2.1 *If $E[\varepsilon^6] < \infty$ and $m^2/\sqrt{N} \to 0$ for $N \to \infty$, then*

$$\mathcal{L}(\sqrt{N}(\hat{\lambda}^*(i,j) - \lambda^*(i,j)) \mid X_1, X_2, \ldots, X_N) \Rightarrow N(0, \tau^2(i,j)) \quad \text{in probability.}$$

The relatively long proof of this theorem which also uses some results about the behaviour of $\hat{F}_N$ as an estimator of F given by Kreiss(1988) is stated in full in Paparoditis(1990) and will not discussed here further. The next section demonstrates the usefulnes of the proposed bootstrap procedure by means of some examples and comparisons.

3   Examples and Comparisons

The proposed Bootstrap procedure which reduces first to innovations which behaves like i.i.d. is not the only way to extend the bootstrap to time dependent stationary observations. Another method has been recently proposed by Künsch(1989). This method operates through randomly choosing overlapping blocks of length l with replacement from the data for which a preliminary fitting of a parametric model is not needed. In order to judge how the proposed Bootstrap procedure works, we estimate the standard error $\hat{\sigma}$ of the lag one correlation coefficient of the AR(1) and MA(1) process in the simulation example considered by Künsch(1989) and compare the results with those reported there. The results of our simulation in which Yule-Walker estimates of the autoregressive parameter, 200 trials and 200 Bootstrap replications have been used, are shown in the following table. An estimate of the standard error of $E(\hat{\sigma}^*)$ is given in parentheses.

First we see that in the AR(1) case, the parametric-bootstrap with m=1 has the smallest mean square error (MSE) as was to be expected. As m increases, the standard error (S.D.) of the estimator increases too. But even in the cases of the overfitted autoregressive models (m=3 and m=9) the standard errors and the mean square errors of the parametric bootstrap are smaller than that reported by Künsch(1989) for the case l=3. The two methods differ also in the bias in that the parametric-bootstrap are always biased upwards. In the MA(1) case the bias of both procedures for m=1 and l=1 is quite similar although the variance of the parametric bootstrap is half times that of the block-bootstrap. Increase of m reduces the bias for both sample sizes considered. Here the standard errors increase too, but for m=3 and m=9 the standard errors of the estimates are quite similar. If we compare the case l=3 with that of m=9 from the point of view of the mean square error, the parametric-bootstrap seems to give better results than the block-bootstrap.

Table 1:   Estimators of the Standard Error of the lag one autocorrelation.

METHOD	AR(1) $\phi = 0.8$ n = 48 E($\hat{\sigma}^*$)	S.D.($\hat{\sigma}^*$)	MSE $\times 10^{-3}$	MA(1) $\theta = 0.8$ n = 48 E($\hat{\sigma}^*$)	S.D.($\hat{\sigma}^*$)	MSE $\times 10^{-3}$	MA(1) $\theta = 0.8$ n = 192 E($\hat{\sigma}^*$)	S.D.($\hat{\sigma}^*$)	MSE $\times 10^{-3}$
1. Parametric-Bootstrap									
m=1	0.114 (0.0007)	0.011	0.17	0.131 (0.0006)	0.008	0.425	0.0638 (0.0003)	0.0035	0.1083
m=3	0.112 (0.0016)	0.022	0.51	0.103 (0.0011)	0.016	0.337	0.0527 (0.0004)	0.0053	0.0298
m=9	0.116 (0.0016)	0.023	0.61	0.111 (0.0012)	0.017	0.290	0.0537 (0.0004)	0.0056	0.0315
2. Block-Bootstrap[+]									
l=1	0.097 (0.0014)	0.020	0.50	0.129 (0.0012)	0.017	0.578	0.0634 (0.0005)	0.0064	0.1293
l=3	0.099 (0.0018)	0.026	0.74	0.106 (0.0012)	0.017	0.325	0.0549 (0.0004)	0.0060	0.0368
3. From 1000 simulations[+]	0.107			0.112			0.0540		

[+] From Künsch(1989).

Consider next the problem of the estimation of the finite sample distribution of the six first partial autocorrelations of the AR(1) and MA(1) model when n=48. The 5% and 95% of the corresponding exact pointwise distribution of $\sqrt{N}(\hat{\lambda}(i,j) - \lambda(i,j))$ obtained by using 2000 replications are plotted in the following figures by a solid line.

Figure 1:   5% and 95% points of the distribution of the first six partial autocorrelations of the AR(1) and MA(1) model.

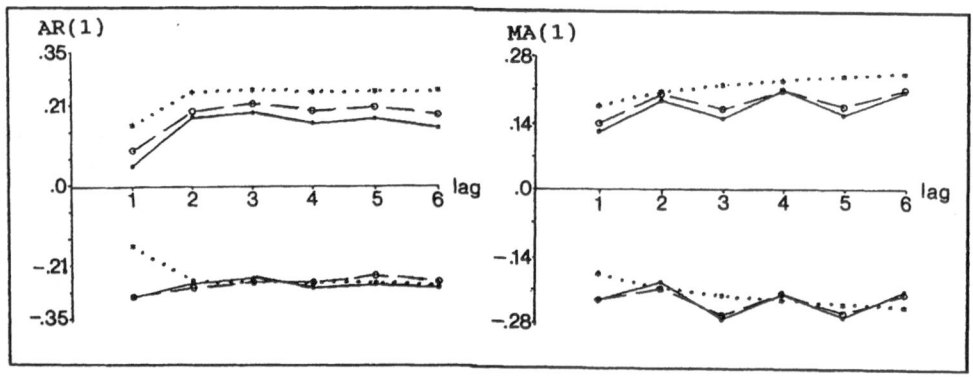

To apply the bootstrap we choose m=11 and B=1000. The bootstrap approximations are shown in this figure by a dashed line. Finaly the corresponding percentage of the asymptotic normal approximation are shown by a dotted line. To estimate the asymptotic variance for each trial, the large sample covariances of the sample autocovariances was estimated using the algorithm given by Mareschal and Mélard(1988). As this figure shows, the boostrap approximates are overall considerably nearer to the points of the exact distribution than the asymptotic normal estimates. It can be also seen, that in the cases in which the exact distribution of the partial autocorrelations is not symmetric, the Bootstrap gives better results than the asymptotic normal approximation.

The next example demonstrates the appropriateness of the proposed bootstrap procedure for the identification of the order p and q of an ARMA model. 200 replications of the ARMA(2,1) process $X_t - 1.2X_{t-1}L + 0.75X_{t-2} = \varepsilon_t + 0.8\varepsilon_{t-1}$ with $\varepsilon_t \sim N(0,1)$ and samples of n=100 and n=200 observations were generated. For each pair (i,j) with $i = 1, \ldots, 6$ and $j = 1, \ldots, 6$ the null hypothesis $\lambda(i,j) = 0$ was tested using the bootstrap distribution estimated by B=1000 bootstrap replications and the significance level 0.05. The entries of the following table show the percentage of cases in which the null hypothesis was rejected in the 200 trials. The pattern implied by the identification property of vector autocorrelations given in section 1 can be clearly seen in these tables.

Table 2: Percentage of cases in which the hypothesis
$\lambda(i,j) = 0$ was rejected for the ARMA(2,1) model.

	N=100						N=200					
j	1	2	3	4	5	6	1	2	3	4	5	6
i=1	100.0	26.0	100.0	99.5	63.0	4.5	100.0	49.5	100.0	100.0	92.5	5.5
2	100.0	100.0	85.0	59.0	32.0	22.0	100.0	100.0	98.5	91.0	73.0	53.0
3	90.5	5.0	4.0	2.0	0.5	0.0	99.5	4.5	5.5	1.5	1.5	1.0
4	68.5	3.5	0.0	0.0	0.0	0.0	93.5	6.5	0.5	0.0	0.0	0.0
5	38.0	2.5	0.0	0.0	0.0	0.0	69.5	1.5	0.5	0.0	0.0	0.0
6	31.0	2.0	0.0	0.0	0.0	0.0	45.5	0.5	0.0	0.0	0.0	0.0

References

Kreiss, J. P. (1988), Asymptotic statistical Inference for a Class of Stochastic Processes, Habilitation work, University of Hamburg.

Künsch, H. R. (1989), "The Jackknife and the Bootstrap for General Stationary Observations," The Annals of Statistics, 17, 1217-1241.

Mareschal, B. and Mélard, G. (1988), "The Corner Method for Identifying Autoregressive Moving Average Models," Applied Statistics, 37, 301-316.

Paparoditis, E. (1990), Vektorautokorrelationen stochastischer Prozeße und die Spezifikation von ARMA-Modellen. Physica-Verlag, Heidelberg.

Streitberg, H. J. (1982), "Vector Correlations of Time Series and the Box-Jenkins Approach to ARMA Identification," Technical Report No. 4/82, Institute of Statistics and Econometrics, Free University of Berlin.

BOOTSTRAP APPROXIMATIONS TO PREDICTION INTERVALS FOR
EXPLOSIVE AR(1)-PROCESSES

W. Stute and B. Gründer
Fachbereich Mathematik, Universität Gießen
Arndtstr. 2, D-6300 Gießen

1. Introduction And Main Result

Let $X_0, X_1, \ldots, X_n$ be observed values from some time series. An important issue then is to predict future values X_{n+s} from the observables. Usually, the quality of the predictor depends on how well a parametric or semiparametric model may be fitted to the data. E.g., if there is strong evidence for an AR(p)-model

$$X_i = \beta_1 X_{i-1} + \ldots + \beta_p X_{i-p} + \varepsilon_i \quad , \tag{1.1}$$

in which the errors $(\varepsilon_i)_i$ are i.i.d. with d.f. F, zero means and finite variance, then the optimal predictor for X_{n+1} under L^2-loss equals

$$\hat{X}_{n+1} = \beta_1 X_n + \ldots + \beta_p X_{n+1-p} \quad .$$

Since usually the regression parameters are unknown, $\beta = (\beta_1, \ldots, \beta_p)$ needs to be replaced by some reasonable estimate $\beta_n = (\hat{\beta}_1, \ldots, \hat{\beta}_p)$:

$$\hat{X}_{n+1} = \hat{\beta}_1 X_n + \ldots + \hat{\beta}_p X_{n+1-p} \quad .$$

In the case of a stationary process, $\beta_n - \beta$ is typically within $O(n^{-1/2})$. The prediction error may thus be decomposed into ε_{n+1} plus a term which is $O(n^{-1/2})$ in probability.

Things change completely if we drop the assumption of stationarity. In this paper we study the simple explosive AR(1)-process

$$X_i = \beta X_{i-1} + \varepsilon_i \quad , \quad 1 \leq i \leq n \quad , \tag{1.2}$$

where $|\beta| > 1$. For estimation of β, we consider the LSE

$$\hat{\beta}_n = \frac{\sum_{i=1}^{n} X_{i-1} X_i}{\sum_{i=1}^{n} X_{i-1}^2} \quad .$$

Interestingly enough, Anderson (1959) showed that $\hat{\beta}_n$ converges to β geometrically fast:

$$\frac{\beta^n}{\beta^2-1} (\hat{\beta}_n - \beta) \longrightarrow Z_\infty^1/Z_\infty^2 \quad \text{in distribution} \quad,$$

where Z_∞^1 and Z_∞^2 are independent copies of

$$Z_\infty = \sum_{i=1}^{\infty} \alpha^{i-1} \varepsilon_i \quad , \quad \alpha = \beta^{-1} \quad .$$

As to the prediction error

$$X_{n+1} - \hat{X}_{n+1} = \varepsilon_{n+1} + (\beta - \hat{\beta}_n) X_n \quad ,$$

however, it turns out that $|X_n|$ tends to ∞ at the same rate as $\hat{\beta}_n$ tends to β, i.e. unlike the stationary case, the second summand is not negligible even for large n.

From a practical point of view, rather than defining a point predictor, a more trustful procedure would be to construct a $1-\rho_1$-prediction interval $\hat{I}_{n,s}$ such that

$$\mathbb{P}(X_{n+s} \in \hat{I}_{n,s} | X_0, X_1, \ldots, X_n) \underset{\sim}{\geq} 1 - \rho_1 \quad .$$

Since the left-hand side is a proper random variable, we are asked to choose $\hat{I}_{n,s}$ such that for a given ρ_2

$$\mathbb{P}\left[\mathbb{P}(X_{n+s} \in \hat{I}_{n,s} | X_0, X_1, \ldots, X_n) \geq 1 - \rho_1\right] \longrightarrow 1 - \rho_2 \quad . \tag{1.3}$$

From a frequentist's point of view, (1.3) states that when e.g. $\rho_2 = 0.1$, 90 in 100 data situations $X_0, X_1, \ldots, X_n$ are such that $\hat{I}_{n,s}$ covers X_{n+s} with probability $\geq 1 - \rho_1$.

A construction of $\hat{I}_{n,s}$ and a proof of (1.3) may be found in Stute and Gründer (1990).

For later reference, it will be outlined below. First, by (1.2),

$$X_{n+s} = \beta^s X_n + \sum_{j=1}^{s} \beta^{s-j} \varepsilon_{n+j} \quad .$$

The interval

$$I_{n,s} = \left[\beta^s X_n + G_s^{-1}(u), \beta^s X_n + G_s^{-1}(1-v)\right]$$

then satisfies

$$\mathbb{P}(X_{n+s} \in I_{n,s} | X_0, X_1, \ldots, X_n) \equiv 1 - \rho_1$$

whenever $u+v = \rho_1$. Here

$$G_s = F * F(\cdot/\beta) * \ldots * F(\cdot/\beta^{s-1})$$

is the d.f. of $\sum_{j=1}^{s} \beta^{s-j} \varepsilon_{n+j}$ (when β is positive). Since in practice, both β and G_s

are unknown, $I_{n,s}$ needs to be replaced by some

$$\hat{I}_{n,s} = \left[\beta_n^s X_n + \hat{G}_{n,s}^{-1}(u), \beta_n^s X_n + \hat{G}_{n,s}^{-1}(1-v) \right] .$$

To define $\hat{G}_{n,s}$, let

$$\hat{\varepsilon}_{ni} = X_i - \hat{\beta}_n X_{i-1} , \quad 1 \le i \le n ,$$

denote the residuals for sample size n, and let $\hat{F}_n$ be their empirical d.f.

(appropriately centered). This serves as a known substitute for the unknown error

d.f. F. For $\hat{G}_{n,s}$ we set (assuming $\beta > 1$ and hence $\hat{\beta}_n > 1$ for n large enough)

$$\hat{G}_{n,s} = \hat{F}_n * \hat{F}_n(\cdot/\hat{\beta}_n) * \ldots * \hat{F}_n(\cdot/\hat{\beta}_n^{s-1}) ,$$

which may be approximated via Monte Carlo. We then showed that

$$||\hat{G}_{n,s} - G_s||_\infty \to 0 \qquad \mathbb{P}\text{-a.s.} \tag{1.4}$$

This has some impact on (1.3), since the conditional coverage probability may be

written as

$$\hat{T}_{n,s}((\hat{\beta}_n^s - \beta^s)X_n) ,$$

where

$$\hat{T}_{n,s}(x) = G_s(x + \hat{G}_{n,s}^{-1}(1-v)) - G_s(x + \hat{G}_{n,s}^{-1}(u))$$

converges, by (1.4), to

$$T_s(x) = G_s(x + G_s^{-1}(1-v)) - G_s(x + G_s^{-1}(u)):$$

$$||\hat{T}_{n,s} - T_s||_\infty \to 0 \qquad \mathbb{P}\text{-a.s.} \tag{1.5}$$

But

$$(\hat{\beta}_n^s - \beta^s)X_n \to c_s Z_\infty \qquad \text{in distribution} ,$$

with $c_s = s\beta^{s-2}(\beta^2-1)$. Hence (under some regularity assumptions on F)

$$\mathbb{P}(X_{n+s} \in \hat{I}_{n,s} | X_0, X_1, \ldots, X_n) \to T_s(c_s Z_\infty) \tag{1.6}$$

by the continuous mapping theorem.

So, for (1.3), we have to choose u and v so that

$$\mathbb{P}(T_s(c_s Z_\infty) \geq 1 - \rho_1) = 1 - \rho_2 \quad .$$

For F unimodal and symmetric at zero, it turns out that for u = v

$$\{T_s(c_s Z_\infty) \geq 1 - \rho_1\} = \{|c_s Z_\infty| \leq w\} \quad .$$

Hence we need to choose u in such a way that w is the $1 - \rho_2$ quantile w_s of H, the d.f. of $|c_s Z_\infty|$. This is guaranteed if

$$T_s(w_s) = 1 - \rho_1 \quad ,$$

i.e.

$$G_s(w_s + G_s^{-1}(1-u)) - G_s(w_s + G_s^{-1}(u)) = 1 - \rho_1 \quad .$$

An approximation for u is obtained by plugging in the empirical approximations of G_s and H. With such a u $\equiv \hat{u}$ it was shown that (1.3) holds.

In a simulation study it was made clear that the approximation (1.3) is pretty good even for moderate sample sizes (n = 40), if ρ_1 and ρ_2 are not too small (as, e.g., 0.01). Though, it may be of some interest to replace the asymptotics by a bootstrap approximation. Generally speaking, estimation of coverage probabilities is an interesting field where the bootstrap may be applied and where the normal competitor does not exist.

The main result is stated as Theorem 1 below. Since typically the arguments are similar to the non-bootstrap situation, we only present the rough ideas. Section 2 presents some simulation results.

To begin with, assume the model (1.2). Put $X_0 = 0$ w.l.o.g. Anderson (1959) showed that

$$\hat{\beta}_n - \beta = \frac{\sum_{i=1}^{n} \varepsilon_i X_{i-1}}{\sum_{i=1}^{n} X_{i-1}^2} \equiv \frac{A_n}{B_n}$$

where A_n and B_n satisfy (in probability)

$$\alpha^{2(n-2)} B_n - \frac{1}{1-\alpha^2} Z_n^2 \to 0 \tag{1.7}$$

$$\alpha^{n-2} A_n - Y_n Z_n \to 0 \quad , \tag{1.8}$$

with

$$Z_n = \sum_{i=1}^{n-1} \alpha^{i-1} \varepsilon_i \qquad\qquad Y_n = \sum_{i=1}^{n} \alpha^{n-i} \varepsilon_i \quad .$$

Observe that

$$Z_n \to Z_\infty = \sum_{i=1}^{\infty} \alpha^{i-1} \varepsilon_i$$

by the martingale convergence theorem. Furthermore, $Y_n = Z_{n+1}$ in distribution. From
(1.7) and (1.8) Anderson (1959) derived the aforementioned fact that

$$\frac{\beta^n}{\beta^2-1} (\hat\beta_n - \beta) \longrightarrow Z_\infty^1/Z_\infty^2 \quad \text{in distribution} . \qquad (1.9)$$

Typically, the limit distribution is long-tailed. On the other hand, the geometric
rate of convergence makes estimation of β extremely good, even for moderate sample
sizes and β as close to 1 as 1.03, say. For the construction of confidence
intervals, (1.9) is of little use, since the limit depends on the whole of F and a
transformation to a pivot is impossible.

In such a situation, the bootstrap offers a useful way to approximate the sampling
distribution of $\hat\beta_n$. The validity of the bootstrap was shown by Basawa et al. (1989).
Resampling is based on the residuals

$$\hat\varepsilon_{ni} = X_i - \hat\beta_n X_{i-1} \quad , \quad 1 \le i \le n \quad .$$

Recall $\hat F_n$, the empirical d.f. of the centered residuals. Generate ε_{ni}^*, $1 \le i \le n$,
independently from $\hat F_n$ and set

$$X_{ni}^* = \beta_n X_{ni-1}^* + \varepsilon_{ni}^* \quad , \quad X_{n0}^* = 0 \quad .$$

The bootstrap value of $\hat\beta_n$ then equals

$$\beta_n^* = \frac{\sum_{i=1}^{n} X_{ni-1}^* X_{ni}^*}{\sum_{i=1}^{n} X_{ni-1}^{*2}} = \beta_n + \frac{A_n^*}{B_n^*} ,$$

say. Basawa et al. (1989) then showed that with probability 1 $\hat\beta_n^n(\beta_n^*-\hat\beta_n)/(\hat\beta_n^2-1)$ has
the same limit as $\beta^n(\hat\beta_n-\beta)/(\beta^2-1)$. So, by a triangle argument, both finite sample
distributions get close together as $n \to \infty$. Similarly, with probability one

$$(\beta_n^{*s} - \hat{\beta}_n^s)X_n^* \to c_s Z_\infty \quad \text{in distribution} \ .$$

Our next concern is with the bootstrap version of $\hat{G}_{n,s}$. Denote with F_n^* the empirical d.f. of $\varepsilon_{n1}^*, \ldots, \varepsilon_{nn}^*$. Set

$$G_{n,s}^* = F_n^* * F_n^*(\cdot/\beta_n^*) * \ldots * F_n^*(\cdot/\beta_n^{*s-1}) \ .$$

By the D-K-W inequality,

$$\mathbb{P}^*(||F_n^* - \hat{F}_n||_\infty \geq \varepsilon) \leq c \exp(-2n\varepsilon^2) \ .$$

Hence, by Borel-Cantelli, with probability one

$$||F_n^* - \hat{F}_n||_\infty \to 0 \qquad \mathbb{P}^*\text{-a.s.}$$

and therefore (use Lemma 2.1 of Stute and Gründer (1990))

$$||F_n^* - F||_\infty \to 0 \ . \tag{1.10}$$

From

$$||A_1 * B_1 - A_2 * B_2||_\infty \leq ||A_1 - A_2||_\infty + ||B_1 - B_2||_\infty$$

we may conclude with probability one

$$||G_{n,s}^* - \hat{G}_{n,s}||_\infty \to 0 \tag{1.11}$$

and therefore

$$||G_{n,s}^* - G_s||_\infty \to 0 \qquad \mathbb{P}\text{-a.s.}$$

The bootstrap analogs of $\hat{T}_{n,s}$ and $\hat{I}_{n,s}$ are given by

$$T_{n,s}^*(x) = \hat{G}_{n,s}(x + G_{n,s}^{*-1}(1-v)) - \hat{G}_{n,s}(x + G_{n,s}^{*-1}(u))$$

$$I_{n,s}^* = \left[\beta_n^{*s} X_n^* + G_{n,s}^{*-1}(u), \ \beta_n^{*s} X_n^* + G_{n,s}^{*-1}(1-v) \right] \ .$$

It is not difficult to see that from (1.11) and (1.4)

$$||T_{n,s}^* - \hat{T}_{n,s}||_\infty \to 0 \ .$$

From (1.5),

$$||T_{n,s}^* - T_s||_\infty \to 0 \ .$$

In summary, we get with probability one

$$\mathbb{P}^*(X_{n+s}^* \in I_{n,s}^* | X_1^*, \ldots, X_n^*) = T_{n,s}^*((\beta_n^{*s} - \hat{\beta}_n^s)X_n^*)$$

$$\to T_s(c_s Z_\infty) \quad \text{in distribution} \ .$$

The bootstrap choice of u is the solution of

$$G_{n,s}^{*}(w_{s}^{*} + G_{n,s}^{*-1}(1-u)) - G_{n,s}^{*}(w_{s}^{*} + G_{n,s}^{*-1}(u)) = 1 - \rho_1 \quad.$$

Note that the left-hand side is non-increasing in u, so that the solution of the above equation is (in principle) unique.

For the computation of w_{s}^{*}, we compute the $1-\rho_2$-quantile of a bootstrap version of $|c_s Z_\infty|$. So, after all, we obtain the following

Theorem 1. Under the same assumptions as for the Theorem in Stute and Gründer (1990), namely

(i)   F is differentiable with $||F'||_\infty < \infty$

(ii) $F'(x) \le c|x|^{-1}$ for all $|x| > K$, some finite c

(iii)   F is symmetric at zero, and F' is unimodal

(iv) F^{-1} is continuous ,

we have with probability one:

$$\mathbb{P}^{*}\left[\mathbb{P}^{*}(X_{n+s}^{*} \in I_{n,s}^{*} \mid X_0^{*}, X_1^{*}, \ldots, X_n^{*}) \ge 1 - \rho_1\right]$$

(1.12)

$$\rightarrow 1 - \rho_2 \quad.$$

So, the left-hand sides of (1.3) and (1.12) have the same limit. In particular, the $\mathbb{P}^{*}$-probability, which is computable via Monte-Carlo, constitutes an approximation for the unknown $\mathbb{P}$-probability in (1.3).

2. Simulations

In the following we present several plots of the exact $(\cdots)$ (computed via Monte-Carlo), bootstrap $(-\cdot-)$ and limit distribution (------) of

$$\mathbb{P}(X_{n+s} \in \hat{I}_{n,s} \mid X_0, X_1, \ldots, X_n) \quad,$$

for different values of ρ, s, n, ρ_1 and ρ_2, with F being the standard normal distribution.

Displays 1 - 4 exhibit that there is almost no difference between the three curves for n as large as n = 200.

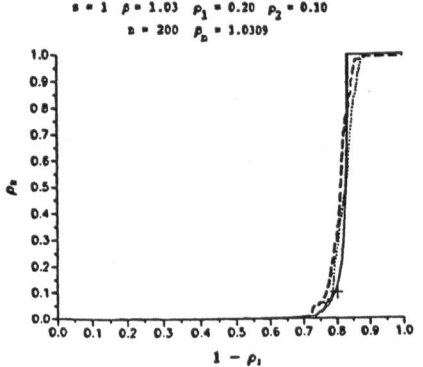

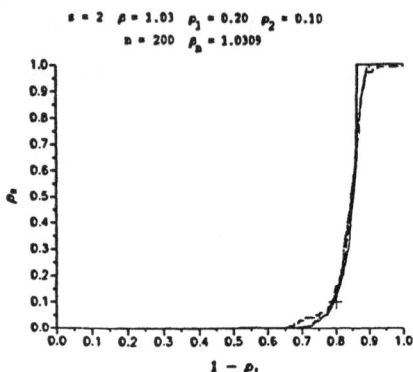

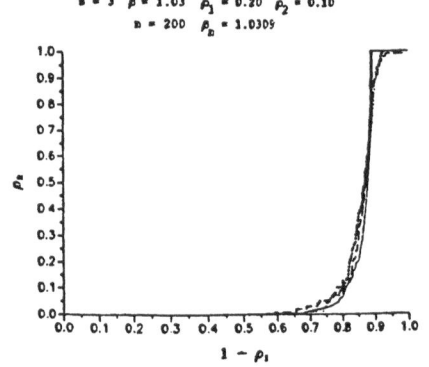

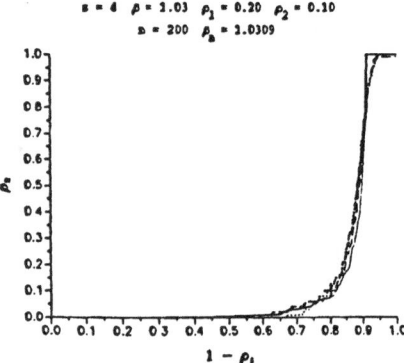

Displays 5 - 8 make clear that the bootstrap outperforms the limit distribution for values of ρ being close to the critical value 1 ($\rho = 1.01$).

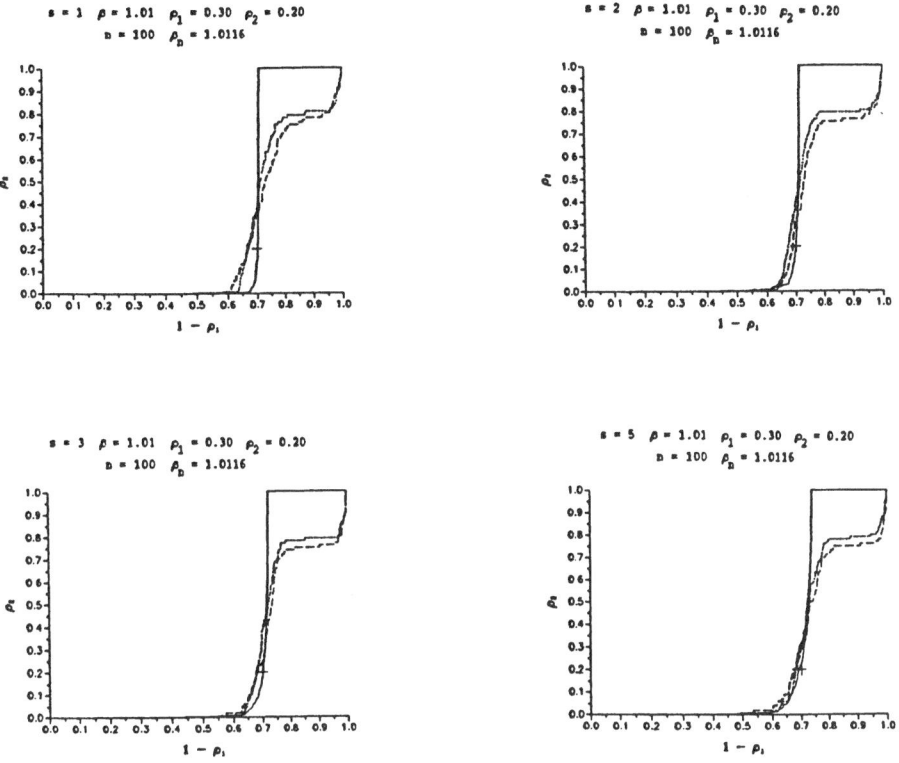

Display 9 is typical for moderate sample sizes (n = 40), with β pretty well estimated by $\hat{\beta}_n$. Finally, the situation underlying display 10 is one of the (very very few) examples where $\hat{\beta}_n$ is a bad estimate of β. As a result, the bootstrap (heavily depending on the $\hat{\beta}_n$, $\hat{G}_{n,s}$, etc.) performs very poor.

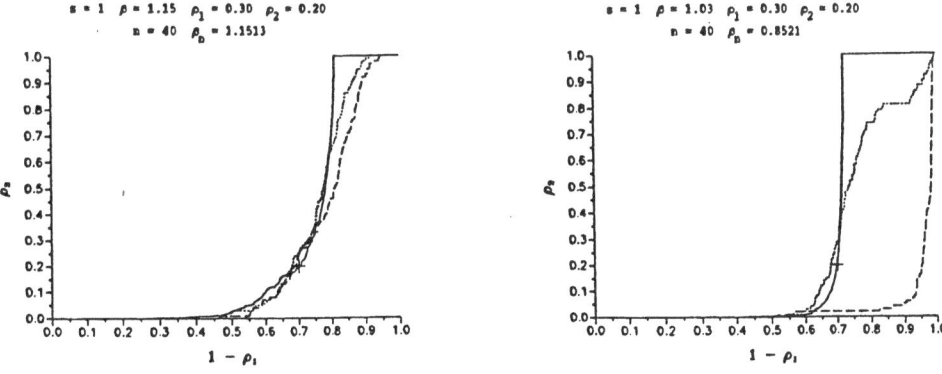

References

Anderson, T.W. (1959). *On asymptotic distributions of estimates of parameters of stochastic difference equations.* Ann. Math. Statist. 30, 676–687.

Basawa, I.V., Mallik, A.K., Mc. Cormick, W.P. and Taylor, R.L. (1989). *Bootstrapping explosive autoregressive processes.* Ann. Statist. 17, 1479–1486.

Stute, W. and Gründer, B. (1990). *Prediction intervals for explosive AR(1)-processes.* Preprint. Univ. of Giessen.

SEARCH FOR A BREAK IN THE PORTUGUESE GDP 1833-1985 WITH BOOTSTRAP METHODS

Isabel Andrade & Isabel Proença [1]
Inst. Sup. Economia e Gestão, Univ. Tecnica Lisboa
Rua Miguel Lupi, 20, 1200 LISBOA, Portugal

1. INTRODUCTION

Since Nelson & Plosser's (1982) seminal paper, the permanent nature of the macroeconomic fluctuations has become the centre of intense debate. The traditional view that macroeconomic time series, namely the real Gross Domestic Product (GDP), are well described as transitory deviations from a deterministic trend is challenged by the identification of a large permanent component in the series, meaning that the fluctuations represent persistent movements, i.e., that the series follow a stochastic trend. The economic implications of this conclusion are substantial, particularly to the Business Cycle Theory (see Andrade (1990)).

Rappoport & Reichlin (1989) and Perron (1989) take a different approach to the study of the nature of the macroeconomic fluctuations by considering the hypothesis of a trend break in the series. The trend break corresponds to a one time innovation with a permanent effect that, when ignored, is confounded with the yearly innovations, making these seem more lasting than they really are. On the contrary, Christiano (1988) shows that there is little statistical evidence against the 'no trend break' null hypothesis for the US post-war real GNP, and that the conventional F critical values have a lack of accuracy that provokes the "excessive" rejection of the null hypothesis. He uses bootstrap methods to calculate a better approximation to the small sample critical values used in the test.

No theoretical results are known for the performance of the bootstrap in this particular problem. However, several authors have proved that, in many situations, the bootstrap approximation has an edge over the classical one, i.e., the bootstrap distribution of the statistic is more accurate then its classic asymptotic distribution (see Babu & Singh (1983), Babu & Bose (1989a), Babu & Bose (1989b), Beran (1982) and Liu & Singh (1987)).

---

[1] We wish to thank Prof. Bento Murteira and two anonymous referees for the helpful comments and encouragement, and Carlos Farinha for technical support. Responsibility for all errors is entirely ours.

In this paper we will study the existence of a trend break in the real Portuguese per capita GDP 1833-1985, the longest time series available for this variable (see Nunes, Mata & Valério (1989), and Figure 1), testing the 'no trend break' null hypothesis for all possible dates. In this way, we are not including any a priori judgements in our analysis.

The plan of this paper is as follows. Section 2 describes the conventional procedure to test the 'no trend break' null hypothesis. Section 3 describes the bootstrap simulation methodology used. Section 4 discusses the results and concludes the paper.

2. CONVENTIONAL PROCEDURE TO TEST THE 'NO TREND BREAK' HYPOTHESIS

As it is well known, there is no agreement on whether the real GDP has a unit root or not, i.e., it is Trend Stationary (TS) or Difference Stationary (DS), using the definitions of Nelson & Plosser (1982). Like Christiano (1988), and to avoid taking sides, we consider both null hypothesis,

$$HO': Y_t = c + \beta t + \phi_1 Y_{t-1} + \phi_2 Y_{t-2} + \epsilon_t \qquad \text{(TS model)}$$

$$HO'': \nabla Y_t = c + \phi \nabla Y_{t-1} + \epsilon_t \qquad \text{(DS model)}$$

The estimated models under the null hypothesis are,

$$HO': \quad \hat{Y}_t = 0.093448 + 0.00060049\,t + 1.0091\,Y_{t-1} - 0.033211\,Y_{t-2} + \hat{\epsilon}_t \quad (TS)$$
$$(0.084332)\ (0.00030645)\ (0.082085)\ \ (0.082363)$$
$$Q(38) = 45.3$$

$$HO'': \quad \nabla\hat{Y}_t = 0.013444 + 0.046695\,\nabla Y_{t-1} + \hat{\epsilon}_t \qquad (DS)$$
$$(0.0065407)\ \ (0.080918)$$
$$Q(38) = 44.8$$

(n=151, with t=-1,0 reserved for initial conditions, and standard deviations under the estimates). Using the Ljung-Box Q statistic, we cannot reject the null hypothesis of serial independence of the residuals for both models.

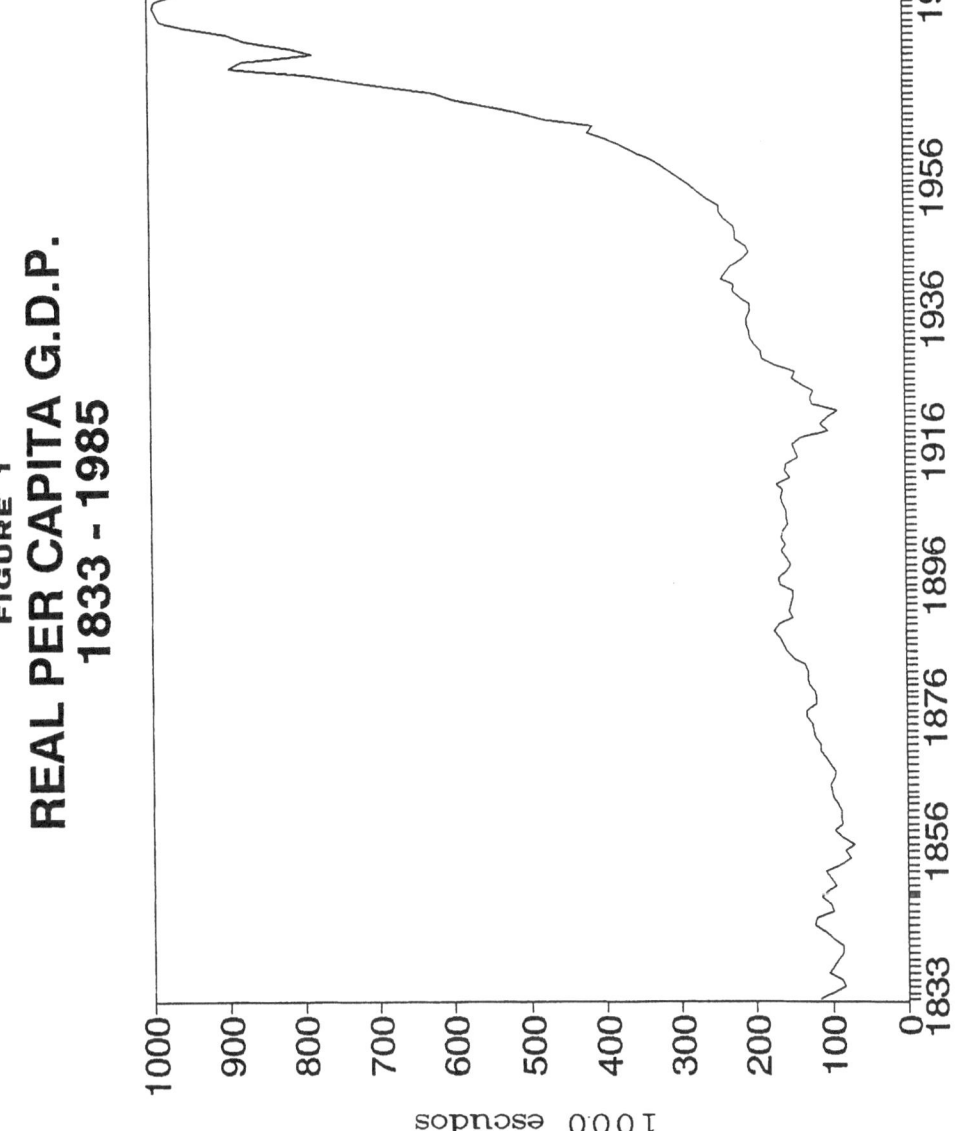

FIGURE 1

REAL PER CAPITA G.D.P.
1833 - 1985

REAL PER CAPITA GDP
1833 - 1985
$(10^3$ escudos$)$

Year	Value	Year	Value	Year	Value
1833	116	1884	149	1935	209
1834	104	1885	158	1936	204
1835	85	1886	162	1937	204
1836	88	1887	167	1938	216
1837	106	1888	176	1939	227
1838	98	1889	168	1940	225
1839	93	1890	151	1941	242
1840	86	1891	154	1942	236
1841	86	1892	153	1943	228
1842	101	1893	151	1944	214
1843	109	1894	150	1945	205
1844	124	1895	166	1946	209
1845	122	1896	169	1947	223
1846	99	1897	158	1948	224
1847	103	1898	152	1949	225
1848	114	1899	157	1950	237
1849	107	1900	164	1951	244
1850	95	1901	160	1952	244
1851	102	1902	165	1953	260
1852	109	1903	163	1954	274
1853	89	1904	156	1955	282
1854	75	1905	157	1956	293
1855	82	1906	158	1957	306
1856	71	1907	162	1958	320
1857	87	1908	164	1959	336
1858	97	1909	163	1960	355
1859	86	1910	170	1961	373
1860	88	1911	152	1962	394
1861	88	1912	159	1963	417
1862	93	1913	157	1964	411
1863	98	1914	142	1965	476
1864	100	1915	145	1966	505
1865	103	1916	148	1967	549
1866	97	1917	136	1968	596
1867	95	1918	102	1969	619
1868	102	1919	113	1970	685
1869	107	1920	103	1971	733
1870	115	1921	90	1972	795
1871	115	1922	123	1973	894
1872	121	1923	126	1974	876
1873	123	1924	121	1975	782
1874	126	1925	137	1976	813
1875	133	1926	148	1977	872
1876	133	1927	145	1978	898
1877	120	1928	174	1979	952
1878	120	1929	187	1980	983
1879	121	1930	189	1981	989
1880	129	1931	199	1982	992
1881	131	1932	204	1983	990
1882	131	1933	206	1984	964
1883	135	1934	209	1985	988

As a first step in looking for trend breaks, we estimate the n-4 regressions to study the possibility of a break at any date i (we illustrate the procedure for the TS model),

$$Y_t = c + \theta d_t^i + \beta t + \gamma d_t^i t + \phi_1 Y_{t-1} + \phi_2 Y_{t-2} + \epsilon_t$$

$$d_t^i = \left\{ \begin{array}{ll} 0 & , \ t = 1, \ldots, i-1 \\ 1 & , \ t = i, \ldots, n \end{array} \right\} \quad \text{for } i = 3, \ldots, n-2$$

where the i'th regression allows both the intercept and the slope to change at date i. Let F_i denote the F statistic for testing $H0': \theta = \gamma = 0$, i.e., that there is no trend break at date i,

$$F_i = \frac{SSR_R - SSR_{Ui}}{r \cdot SSR_{Ui}/(n-k)} \quad (i = 1, 2, \ldots, n-1)$$

where SSR is the sum of the squared residuals, the index R denotes the model under H0 (Restricted model), the index U denotes the model under H1 (Unrestricted model), r is the number of restrictions (r=2), and k is the number of parameters of the model under H1 (k=6). This statistic has an asymptotic F distribution with (r,n-k) degrees of freedom.

The conventional procedure to test a trend break is to estimate the Restricted and Unrestricted models for one possible trend break at date i, chosen using a priori information, and test the F statistic against the upper 5% critical value of the F distribution. Here, we calculate the complete sequence of F statistics for all the possible trend break dates (where the Restricted model is the same for all i), and then, following Christiano (1988) in order to avoid the multiple test problem, we select the locally maximal F statistics and 'test' them against the upper 5% critical value of the F distribution [2].

For the TS model, we obtained rejections of the null hypothesis, i.e., 'statistically significant' trend breaks, in the years 1838, 1889, 1897, 1911, 1914, 1917, 1953, and 1965 (Figure 2). For the DS model, we obtained rejections of the null hypothesis in the years 1857, 1922, 1947, 1950 and 1953 (Figure 3), although the general pattern of the F_i statistic

---

[2] This 5% critical value does not correspond to an overall significance level for the complete sequence of the tests, but only to the level one investigator would have calculated if he had chosen to test that particular date. This same interpretation is used by Prof. Hendry in the PC-GIVE package (see Hendry (1989)).

plot is similar for both models. Most unexpectedly, no rejection of H0 was found in the period 1973-1975, which corresponds to the first oil price shock, and the change of political regime in Portugal.

We find excessive the number of rejections of the 'no trend break' null hypothesis for both models, particularly if we look at the values of the F_i statistic for all possible dates of break, and not only at the locally maximal F_i. Therefore, we looked for a better approximation to the distribution of the F_i statistic, and, following the paper of Christiano (1988) for the US GNP, we chose the bootstrap method. This choice is well justified by the good performance of the bootstrap in related econometric problems: Freedman (1981, 1984), and Freedman & Peters (1984) have established the validity of the bootstrap for some regression problems, and Bose (1988) showed that for autoregression models the bootstrap distribution of the estimators is more accurate than the classic asymptotic normal distribution.

3. BOOTSTRAP SIMULATION METHODOLOGY

Since we accept the hypothesis of serial independence of the residuals, the bootstrap was performed with resampling from the empirical distribution of the fitted residuals, according to the following algorithm:

1) Draw randomly n times with replacement one observation from the set of fitted residuals of the Restricted model. This procedure generates a set of n bootstrap artificial residuals, $\hat{\epsilon}_t^*$ and $\hat{\epsilon}_t'^*$ $(t = 1, 2, ..., n)$;

2) Using the bootstrap residuals, construct the artificial data for the endogenous variables according to,

$$Y_t^* = \hat{c} + \hat{\beta} t + \hat{\phi}_1 Y_{t-1}^* + \hat{\phi}_2 Y_{t-2}^* + \hat{\epsilon}_t^* \qquad \text{(TS model)},$$

and,

$$\nabla Y_t^* = \hat{c}' + \hat{\phi}' \nabla Y_{t-1}^* + \hat{\epsilon}'_t^* \qquad \text{(DS model)},$$

considering $Y_0^* = Y_0$ and $Y_{-1}^* = Y_{-1}$ for t=1,2,...,n. The coefficients c, $\hat{\beta}$, $\hat{\phi}_1$, $\hat{\phi}_2$, c', and $\hat{\phi}'$ are the OLS estimates previously obtained;

3) With the artificial data, estimate the Restricted model, and obtain SSR_R^*. For each null hypothesis, estimate the Unrestricted model (which includes the i'th dummy variable d_t^i), and obtain the sum of squared residuals SSR_{Ui}^*;

4) For each null hypothesis, calculate the statistic,

$$F_i^* = \frac{SSR_R^* - SSR_{Ui}^*}{r \cdot SSR_{Ui}^*/(n-k)} \qquad (i = 1, 2, \ldots, n-4) \quad ;$$

5) Repeat 1 to 4 a sufficiently large B number of times. In this paper we used B=1000.

The critical values for each test (i=1,2,...,n-4) at the significance level of 5% belong to the upper 5% percentile of the empirical bootstrap distribution of the F_i^* statistic [3] .

4. RESULTS

Using the previous algorithm, we estimated the upper 5% bootstrap critical values for each of the models (plotted in Figures 2 and 3). We used the TSP package (Time Series Processor) in a MicroVAX 3600. It uses the Multiplicative Congruential Method with modulus $2^{31}-1$ and multiplier 397204094 to generate the random numbers (see Fishman & Moore (1982)).

For the TS model, the bootstrap upper 5% values show a curious pattern (already found in Christiano's 1988 paper), revealing that the bootstrap distribution slowly 'moves' to the right and, after attaining the middle point, moves back. On the contrary, for the DS model, the bootstrap upper 5% critical values are approximately the same for all potential trend break dates. This situation is similar to what happens with the conventional F distribution (same upper 5% critical value for all dates).

We can see from Figures 2 and 3 that the bootstrap upper 5% critical values are higher than the upper 5% values of the conventional F distribution, allowing us to reduce the number of rejections of the null hypothesis in both models. For the TS model, we only have a significant trend break in 1914, whereas for the DS model we have two, 1922 and 1947. These new results are coherent with the accepted analysis of the Portuguese Economy (the exception is the 1973-75 period already mentioned), thus indicating the poor quality of the F approximation conventionally used in the test of the trend breaks, and the improvements we can obtain by using bootstrap methods.

REFERENCES
Andrade I. (1990). "Tendência e Ciclo no Produto Português - Uma contribuição para o teste da Teoria dos Ciclos Económicos Reais em Portugal", Unpublished M.Sc. dissertation, ISEG, Lisboa.

---

3 See note 2.

FIGURE 2

EMPIRICAL F STATISTICS AND BOOTSTRAP CRITICAL VALUES (TS MODEL)

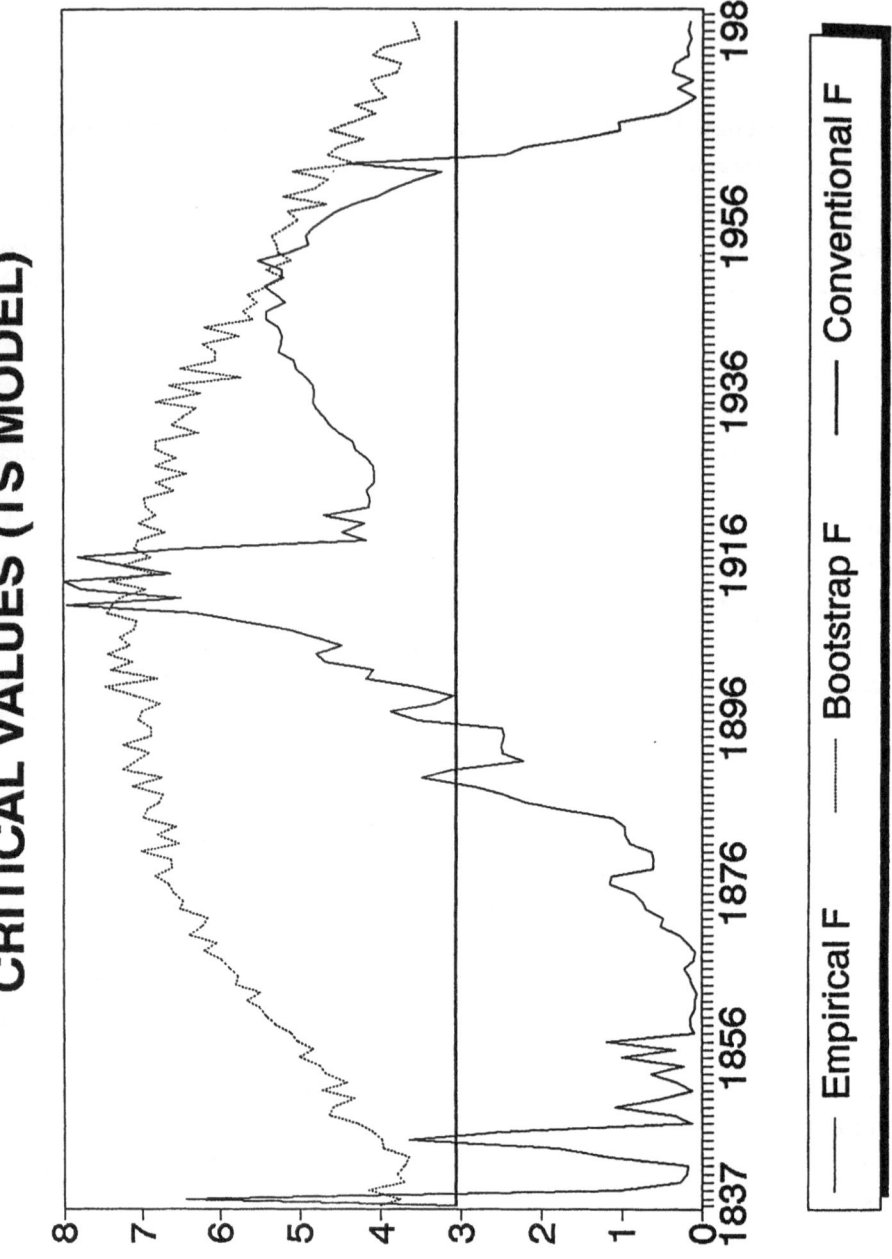

FIGURE 3

EMPIRICAL F STATISTICS AND BOOTSTRAP CRITICAL VALUES (DS MODEL)

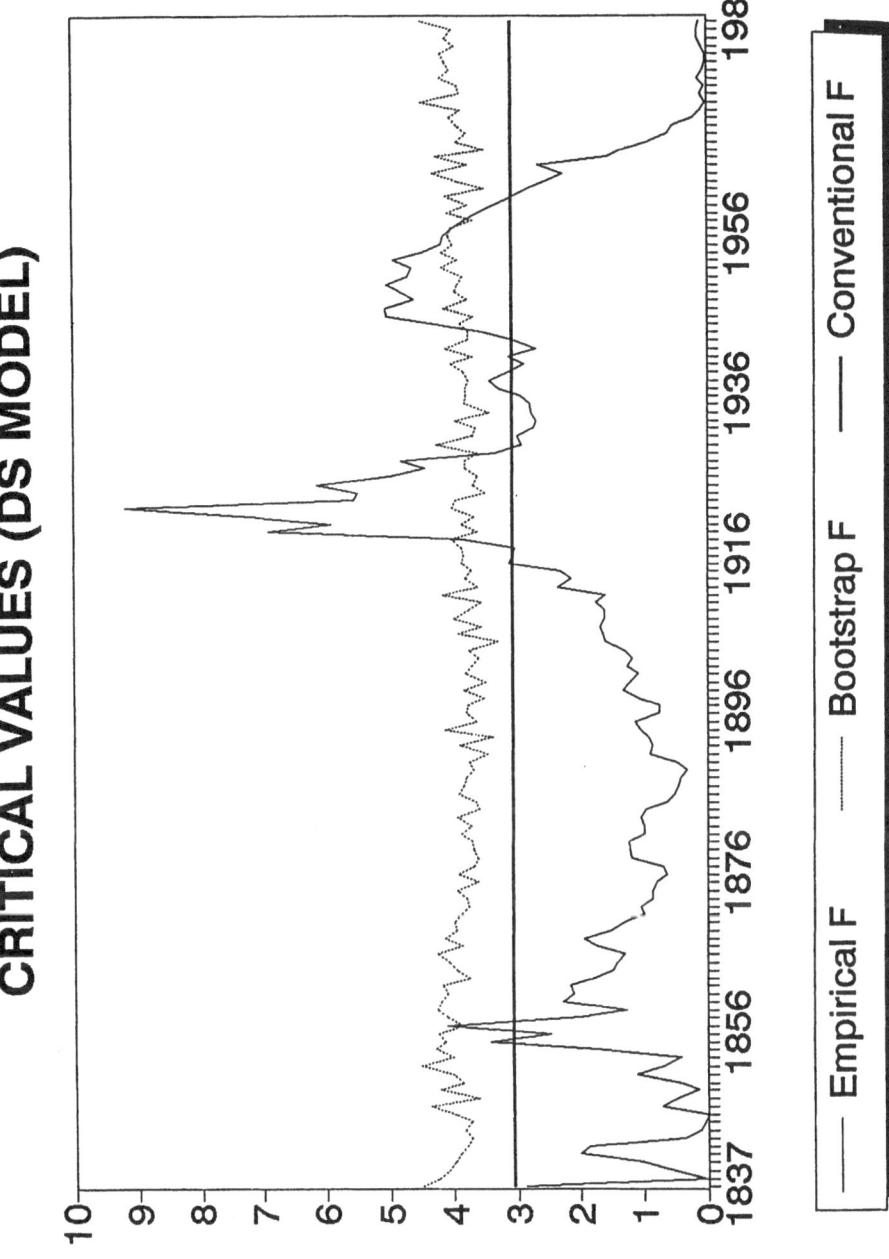

Babu, G.J.; Bose A. (1989a). "Bootstrap Confidence Intervals", Statist. Probab. Let., 7, 151-160.

Babu, G.J.; Bose A. (1989b). "Accuracy of the Bootstrap Approximation", Purdue University, Technical Report #89-10.

Babu, G.J.; Singh, K. (1983). "Inference on Means Using the Bootstrap", Annals Statist., 11, 999-1003.

Beran, R. (1982). "Estimated Sampling Distributions: The Bootstrap and Competitors", Annals Statist., 10, 212-225.

Bose, A. (1988). "Edgeworth Correction by Bootstrap in Autoregressions", Annals Statist., 16, 1709-1722.

Christiano,L.J. (1988). "Searching for a Break in GNP", NBER, Working Paper #2965.

Efron, B. (1979). "Bootstrap Methods: Another Look at the Jackknife", Annals Statist., 7, 1-26.

Efron, B.; Gong, G. (1983). "A Leisurely Look at the Bootstrap, the Jackknife, and Cross-Validation", The American Statistician, 37, 36-48.

Fishman,G.S.; Moore,L.R. (1982). "A Statistical Evaluation of Multiplicative Congruential Random Number Generators with Modulus $2^{31}-1$", JASA, 77, 129-136.

Freedman, D.A. (1981). "Bootstrapping Regression Models", Annals Statist., 9, 1218-1228.

Freedman, D.A. (1984). "On Boostrapping Two-Stages Least-Squares Estimates in Stationary Linear Models", Annals Statist., 12, 827-842.

Freedman, D.; Peters, S.C. (1984). "Bootstrapping a Regression Equation: Some Empirical Results", JASA, 79, 97-106.

Hendry, D.F. (1989). "PC-GIVE: An Interactive Econometric Modelling System", Institute of Economics and Statistics, Oxford.

Liu, R.Y.; Singh, K. (1987). "On a Partial Correction by The Bootstrap", Annals Statist., 15, 1713-1718.

Nelson,C.R.; Plosser,C.I. (1982). "Trends and Random Walks in Macroeconomic Time Series", J. Monet. Econ., 10, 139-162.

Nunes,A.B.; Mata,E.; Valério,N. (1989). "Portuguese Economic Growth 1833-1985", J. Europ. Econ. History, 18, 291-330.

Perron,P. (1989). "The Great Crash, the Oil Price Shock, and the Unit Root Hypothesis", Econometrica, 57, 1361-1401.

Proença,I. (1990). "Método Bootstrap - Uma Aplicação na Estimação e Previsão em Modelos Dinâmicos", Unpublished M.Sc. dissertation, ISEG, Lisboa.

Rappoport,P.; Reichlin,L. (1989). "Segmented Trends and Non-stationary Time Series", Econ. Journal, 99, 168-177.

ONE-STEP BOOTSTRAPPING IN GENERALIZED LINEAR MODELS

Olaf Mosbach

FB 03: Institut für Statistik, Universität Bremen

Postfach 330 440, D-2800 Bremen 33

1. Introduction:

Some very commonly used statistical models are members of the class of generalized linear models introduced by Nelder & Wedderburn(1972). Included are probit and logit models for binomially distributed data, log-linear models for Poisson and also classical linear models for normally distributed response data. In all these examples the underlying distribution is an element of an one-parametric exponential family. The ML-estimator for the mean vector has an interpretation as projection of the observation vector onto the space of all possible mean values. The computation is in most situations an iteratively weighted least-square estimation. However, existence and uniqueness of this estimator is not always guaranteed (see Wedderburn(1976) for a set of sufficient conditions). The following example presents logistic regression with explanatory variables x_n and only 10 0-1-observations Y_n. Here the probability of the existence of the ML-estimator $\hat{\beta}(Y_1,...,Y_{10})$ is only around 80%.

Example: $Y_1, ... ,Y_{10}$ independent distributed randomvariables with:

$\mathscr{L}(Y_n) = $ Binomial$(1, \mu_n)$, $\mu_n = $ logit$^{-1}(\beta_0 + \beta_1 x_n)$, n=1, ... ,10

$$\beta = (\beta_0, \beta_1)^T = (3.1852, -0,4055)^T$$

n	1	2	3	4	5	6	7	8	9	10	
x_n	5	8	9	10	10	10	11	13	14	14	
μ_n	76	49	39	30	30	30	22	11	8	8	[%]

The data consists of 10 observations randomly selected from a real problem published by Rothe(1986). The estimated parameter is assumed to be true.

Even if the estimator exists other computational problems can arise. The results of the iteration depend on the choice of a starting value, the location of all estimators (if not unique) and on the number of iteration steps. So every bootstrap method in generalized linear models must rebuild a possibly unstable estimator. Moulton & Zeger(1985) proposed to simplify the bootstrap estimation by an one-step method: Take the usual estimator as starting value, resample new observations and iterate one step to the bootstrap estimator. In the following sections this idea will be interpreted geometrically and linear solutions will be developed for parametric and non-parametric resampling. The proposed bootstrap estimators are of a simple form and have the same asymptotic properties as the estimator above.

2. Generalized Linear Models

The data for a single observation (Y_n, X_n, φ_n) consist of a real randomvariable Y_n, an associated vector of covariables $X_n \in \mathbb{R}^P$ and a weight φ_n, if necessary. The mean μ_n of the response Y_n is related to the explanatory variables X_n through a linear predictor $\eta_n = X_n^T \beta$, $\beta \in \mathbb{R}^P$ and a known link function g:

$$E(Y_n) = \mu_n = g(\eta_n) = g(X_n^T \beta) \quad \text{with g twice continous differentiable} \tag{1}$$
$$\text{and positive first derivation: } g' > 0$$

For a geometric definition of the estimator no distributional assumptions are necessary. It suffices to know a variance function v, that connects the variance of Y_n with the mean:

$$\text{Var}(Y_n) = v(\mu_n, \varphi_n) \quad \text{with v continous in } \varphi_n, \text{ cont.differentiable in } \mu_n, \text{ v} > 0 \tag{2}$$

Consider a sample of N independent observations $Y = (Y_1, ..., Y_N)^T$ with realization y, $X = (X_1, ..., X_N)^T$ and $\varphi = (\varphi_1, ..., \varphi_N)^T$. Then the estimator $\hat{\mu}$ of $\mu = (\mu_1, ..., \mu_N)^T$ can be regarded as a projection of y onto the non-linear space of all possible mean vectors $\left\{ \left[g(X_n^T \xi) \right]_{n=1,...,N} \mid \xi \in \mathbb{R}^P \right\}$. The projection depends on the variance structure, that is possibly different at each point of the surface. In other words: The difference between the observation y and the estimated mean $\hat{\mu}$ is, given the estimated variance structure $V(\hat{\mu}) = \text{diag}\left(\left[v(\hat{\mu}_n, \varphi_n) \right]_{n=1,...,N} \right)$, orthogonal to the tangent space $\left\{ G'(\hat{\eta})X\xi \mid \xi \in \mathbb{R}^P \right\}$, $G'(\hat{\eta}) = \text{diag}\left(\left[g'(\hat{\eta}_n) \right]_{n=1,...,N} \right)$ at the estimated point of the surface:

$$0 = X\, G'(\hat{\eta})\, V^{-1}(\hat{\mu})\, (y - \hat{\mu}) \tag{3}$$
$$= \hat{T}^T \hat{r} \qquad \text{with} \quad \hat{\eta} = g^{-1}(\hat{\mu}), \quad \hat{\beta} = X^- \hat{\eta}, \text{ if rank(X)=P}$$
$$\hat{T} = V^{-1/2}(\hat{\mu})\, G'(\hat{\eta})\, X$$
$$\hat{r} = V^{-1/2}(\hat{\mu})\, (y-\hat{\mu}) \quad \text{`Pearson residuals'}$$

McCullagh & Nelder(1989) call this way quasi-likelihood estimation. In fact if $\mathfrak{L}(Y_n)$ is an element of an one-parametric exponential family, it coincides with the ML-estimator. The Newton-Raphson algorithm computes $\hat{\beta}$ by stepwise correction of a starting value $\hat{\beta}^{(O)}$:

$$\hat{\beta}^{(t+1)} = \hat{\beta}^{(t)} + \hat{F}^{(t)-1}\, \hat{T}^{(t)T}\, \hat{r}^{(t)} \xrightarrow{\ t \to \infty\ } \hat{\beta} \tag{4}$$
$$\text{with:} \quad T(\beta) = V^{-1/2}(\mu)\, G'(\eta)\, X, \quad \hat{T}^{(t)} = T(\hat{\beta}^{(t)})$$
$$F(\beta) = T^T(\beta)\, T(\beta), \quad \hat{F}^{(t)} = F(\hat{\beta}^{(t)})$$
$$r(\beta) = V^{-1/2}(\mu)\, (y-\mu), \quad \hat{r}^{(t)} = r(\hat{\beta}^{(t)})$$

The one-step bootstrap will fix the tangent space of the last iterated point $\hat{\mu}$ and estimates the variability of $\hat{\beta}$ around the true value β with the correction of $\hat{\beta}$ under a new resampled observation vector:

$$(\beta - \hat{\beta}) \simeq (\beta^* - \hat{\beta}) = \hat{F}^{-1} \hat{T}^T R^* \quad \text{with resampling } R^* \tag{5}$$

Now the bootstrap approximation is just a linear function of the resampling and therefore protected against any computational problems.

3. Normal Approximation

Assume the number of observations N tends to infinity. Suppose that the regularity conditions: – bounded covariables: $X_n \in \mathfrak{X}$ compact, $n \in \mathbb{N}$
  – bounded weights: $\varphi_n \in \Phi$ compact, $n \in \mathbb{N}$
  – the 3rd. and 4th. moments of Y_n are continous in (μ_n, φ_n)
  – convergence of F: $(1/N)\, F(\beta) \xrightarrow{\ N \to \infty\ } \Sigma$ positive definite   hold.

Then $\hat{\beta}$ asymptotically exists with probability 1,
is consistent: $\hat{\beta}(N) \xrightarrow{\ N \to \infty\ } \beta$ a.s. ,
and asymptotical normally distributed: $\sqrt{N}\, (\hat{\beta} - \beta) \xrightarrow{\ \mathfrak{L}\ } N(0, \Sigma^{-1})$.

The proof can be shown by similar arguments as in Fahrmeir & Kaufmann(1985) even under some weaker assumptions.

This leads to the normal approximation: $(\beta - \hat{\beta}) \overset{\mathfrak{L}}{\approx} N(0, \hat{F}^{-1})$

$$(\beta - \hat{\beta}) \approx \hat{F}^{-1/2} R^* = \hat{F}^{-1} \hat{F}^{1/2} R^* \text{ with } \mathfrak{L}(R^*) = N_p(0, \mathbb{1}), \tag{6}$$

that has an analogous form as the one-step bootstrap (5) with normally distributed `resampling´. But in finite situations with very small samples the normal approximation fails:

> ### Example (cont.):
> In this example, the exact confidence level of symmetrically constructed 95% confidence intervals for β_1 is only 53.1%, provided $\hat{\beta}$ exists.

This is the operational area of the one-step bootstrap: The resampling of a small number of observations shows good results, because even a moderate number of bootstrap replications is comparable with the number of all possible observations.

4. Parametric Bootstrap

Assume a parametric model like an exponential family, with:

$$\mathfrak{L}(Y_n) = \Pi(\mu_n \mid \varphi_n) \qquad \Pi \text{ one-parametric distribution family } (\varphi_n \text{ fixed}).$$

Then it is possible to simulate new bootstrap replications Y* of Y by:

$$Y^* = (Y_1^*, ..., Y_N^*)^T \quad \text{with} \quad \mathfrak{L}(Y_n^*) = \Pi(\hat{\mu}_n \mid \varphi_n) \quad n=1,...,N \quad \text{independent;}$$

and treat these like ordinary observations in an one-step iteration from starting point $\hat{\beta}$. Hence R* is equal to Y_n^* standardized with the known mean and variance:

$$R^* = V^{-1}(\hat{\mu})(Y^* - \hat{\mu}) = \overline{Y^*}$$

and $\quad (\beta - \hat{\beta}) \simeq (\beta^* - \hat{\beta}) = \hat{F}^{-1}\hat{T}^T\,\overline{Y^*}$. $\hfill (7)$

Under the assumptions of section 3, especially $\hat{\beta}(N) \xrightarrow{N\to\infty} \beta$ a.s., one shows:

$$E^*(\sqrt{N}(\beta^* - \hat{\beta})) = 0,$$
$$Var^*(\sqrt{N}(\beta^* - \hat{\beta})) = N\,\hat{F}^{-1} \xrightarrow{N\to\infty} \Sigma^{-1} \text{ a.s.}$$

and: $\quad \sqrt{N}(\beta^* - \hat{\beta}) \xrightarrow{\mathfrak{L}} N(0, \Sigma^{-1}) \quad$ a.s. ,

because the 4th moments of $\left\{v^T\,\hat{F}^{-1/2}\,\hat{T}_n^T\,\overline{Y_n^*}\,,\ n\in\mathbb{N}\right\}$ are bounded for all $\|v\|=1$. The asymptotic normality is also valid for complete iterations, as shown by Rothe(1986).

5. Non-parametric Bootstrap

The problem of resampling from an empirical distribution is to find some identical distributed observations. Moulton & Zeger(1985) suggested to resample the data as a whole:

$$\mathfrak{L}((Y_n^*, X_n^*, \varphi_n^*)) = \text{Empirical}\left\{(Y_n, X_n, \varphi_n),\ n = 1, ... , N\right\} \quad n=1,...,N \quad \text{independent}$$

and got after an one-step iteration with starting value $\hat{\beta}$:

$$(\beta^* - \hat{\beta}) = (\hat{T}^T\,\text{diag}(D^*)\,\hat{T})^{-1}\hat{T}^T\,\text{diag}(D^*)\,\hat{r}.$$

Here D* denotes a multinomial distributed drawing indicator:

$$\mathfrak{L}(D^*) = \text{Multinomial}(N, 1/N\ (1,...,1)).$$

A linearisation at the expected value of D* gives the one-step bootstrap:

$$(\beta - \hat{\beta}) \simeq (\beta^* - \hat{\beta}) = \hat{F}^{-1}\hat{T}^T\,\text{diag}(D^*)\,\hat{r}$$
$$= \hat{F}^{-1}\hat{T}^T\,\text{diag}(\hat{r})\,D^* = \hat{F}^{-1}\hat{T}^T R^* \hfill (8)$$

with new weighted Pearson residuals as resampling:

$$R^* = \text{diag}(\hat{r})\,D^*.$$

Assume additional to the assumptions of section 3:

$\quad - \hat{T}^T\hat{r} = 0, \qquad$ - bounded observations: $y_n \in \Upsilon$ compact, $n\in\mathbb{N}$.

It can be shown that

$$E^*(\sqrt{N}(\beta^* - \hat{\beta})) = 0$$

$$Var^*(\sqrt{N}(\beta^* - \hat{\beta})) = N \, \hat{F}^{-1} (\hat{T}^T \, \text{diag}^2(\hat{r}) \, \hat{T}) \, \hat{F}^{-1}$$

$$\xrightarrow{N \to \infty} \Sigma^{-1} \quad \text{a.s.}$$

$$\sqrt{N}(\beta^* - \hat{\beta}) \xrightarrow{\mathfrak{L}} N(0, \Sigma^{-1}) \quad \text{a.s.} ,$$

with the strong law of large numbers and a theorem of Morris(1975, lemma2.2).

Example (cont.):

In this example, the distribution of the bootstrapped parameter β_1^* is approximated by its empirical distribution after 100 respectively 500 parametric and non-parametric resamples. The exact confidence levels of symmetrical 95% confidence intervals for β_1 via β_1^* are:

number of bs-replications	parametric	non-parametric
100	99.4%	88.4%
500	99.0%	91.1%

provided $\hat{\beta}$ exists.

Thus the one-step bootstrap works in situations, where asymptotic results are useless. The computation is only a linear function of a random component and allows a great number of resampled observations. Only if the number of observations requires too much bootstrap resamples, bootstrap fails, but then the normal approximation works.

REFERENCES:

Fahrmeir, L. & Kaufmann, H. (1985). Consistency and Asymptotic Normality of the ML Estimator in Generalized Linear Models. *Annals of Statistics* **9**, pp.342-368. (Correction: **14**, p.1643).

McCullagh, P. & Nelder, J.A. (1989). *Generalized Linear Models*, 2nd. Edition. London, New York: Chapman and Hall.

Morris, C. (1975). Central Limit Theorems for Multinomial Sums. *Annals of Statistics* **3**, pp.165-188.

Mosbach, O. (1988). *Bootstrap-Verfahren in Allgemeinen Linearen Modellen*. Universität Bremen, Fachbereich 03: Diplomarbeit.

Moulton, L.H. & Zeger, S.L. (1985). Bootstrapping Generalized Linear Models. Baltimore, Johns Hopkins University: *Department of Biostatistics Paper* **No.572.**

Nelder, J.A. & Wedderburn, R.W.M. (1972). Generalized Linear Models. *Journal of the Royal Statistical Society, Series A* **135**, pp.370-384

Rothe, G. (1986). Bootstrap in Generalisierten Linearen Modellen. Mannheim: *ZUMA-Arbeitsbericht* **Nr.86/11**

Wedderburn, R.W.M. (1976). On the Existence and Uniqueness of the ML-Estimates for certain Generalized Linear Models. *Biometrika* **63**, pp.27-32.

BOOTSTRAP FOR MEAN AND COVARIANCE STRUCTURE MODELS

Günter Rothe
Hauptverband der gewerblichen
Berufsgenossenschaften
Referat Statistik
Alte Heerstr. 111
D–5202 St. Augustin 2
Federal Republic of Germany

Gerhard Arminger
Bergische Universität
Fachbereich 6
Gaußstr. 20
D–5600 Wuppertal
Federal Republic of Germany

1   Introduction

The development of the social sciences as empirical sciences has been hampered by their great difficulties to measure variables of substantive interest to the researcher using reliable and valid instruments. Consequently, questions of scaling and measurement have been of great theoretical and practical concern. Researchers working in areas such as sociology, economics, and epidemiology usually wish to connect scales that measure some dependent variables with each other and with explanatory variables. Hence, models that incorporate measurement models and regression models simultaneously have been often applied in these areas of research. The most popular one is the LISREL model (cf. Jöreskog and Sörbom, 1988). A somewhat more general model is the following one:

$Y_i, i = 1, \ldots, n \in \mathcal{N}$ are independent, identically distributed random $\mathcal{R}^p$-vectors, where each Y_i is distributed as

$$y = \Lambda \eta + \epsilon \qquad \text{(measurement model for } y\text{)}, \tag{1}$$

and where

$$\eta = B\eta + \zeta \qquad \text{(structural model)}. \tag{2}$$

Here ζ and ϵ are independent random vectors with

$$E(\zeta) = 0, \ var(\zeta) = \Psi, \ E(\epsilon) = 0, \ var(\epsilon) = \Theta \ \ . \tag{3}$$

B and Λ are denoted as "structural", and Ψ and Θ as "covariance parameters". The parameter space is usually restricted to a linear subspace according to the model under consideration; the rest of the parameters has to be estimated from the sample $Y_i, i = 1, \ldots, n$. Usually, this will be done by minimizing a discrepancy function between the empirical covariance matrix

$$S = \frac{1}{n} \sum_{i=1}^{n} Y_i Y_i' \tag{4}$$

and the covariance implied by the model above:

$$\Sigma = \Lambda(I - B)^{-1}\Psi((I - B)^{-1})'\Lambda' + \Theta \quad , \tag{5}$$

where I denotes the identity matrix; here Σ is a function of the $q \leq min[n, p(p+1)/2]$ free parameters of Λ, B, Ψ, and Θ, which can be summarized into a parameter vector $\delta \in \mathcal{M}$, where $\mathcal{M}$ is some open subset of $\mathcal{R}^q$. For all estimation methods it is assumed that δ is second order identifiable, i.e., $\Sigma(\delta_1) = \Sigma(\delta_2)$ implies $\delta_1 = \delta_2$.

The model outlined has been generalized in many ways to include means $\mu(\delta)$, multiple levels of latent variables, non-linear relations, censored metrics as well as dichotomous and ordered variables. In general, models of the form $E(y) = \mu(\delta)$, $var(y) = \Sigma(\delta)$ are considered as *mean and covariance structures*. For simplicity, we restrict ourselves to the case of the model indicated above.

2 Estimation of covariance structures

We focus on the $\mathcal{R}^q$-Parameter δ of a covariance structure $\Sigma(\delta)$ of the $\mathcal{R}^p$-vector Y: if Y_i, $i \in N$ are multivariate normal $(Y_i \sim \mathcal{N}(0, \Sigma(\delta)))$, the maximum likelihood estimator (MLE) maximizes the log-likelihood function

$$l_n(\delta) = \frac{1}{n}\sum_{i=1}^{n} \log f(Y_i; \delta) \tag{6}$$

with

$$\log f(y; \delta) = -\frac{1}{2}[\log \det \Sigma(\delta) + \operatorname{tr}(yy')\Sigma(\delta)^{-1}] \quad . \tag{7}$$

Hence,

$$l_n(\delta) = -\frac{1}{2}[\log \det \Sigma(\delta) + \operatorname{tr}S\Sigma(\delta)^{-1}] \quad . \tag{8}$$

Under $\delta := \delta_0$, the ML estimator $\hat{\delta}_n$ is consistent and asymptotically normal, i.e. $\sqrt{n}(\hat{\delta} - \delta_0) \longrightarrow \mathcal{N}(0, V(\delta_0))$, where $V(\delta_0)$ is the inverse of the Fisher information matrix. Since

$$V(\delta_0) = (-A(\delta_0))^{-1} \quad , \tag{9}$$

where $A(\delta)$ is the expected value of the second derivative of the log-likelihood function, $V(\delta_0)$ as well as $var(\hat{\delta})$ can be estimated using $\hat{A} = A_n = (a_{jk})_n$, where

$$a_{jk} = -\frac{1}{2}\operatorname{tr}[\Sigma^{-1}(2S - \Sigma)\Sigma^{-1}\frac{\partial \Sigma}{\partial \delta_j}\Sigma^{-1}\frac{\partial \Sigma}{\partial \delta_k} - \Sigma^{-1}(S - \Sigma)\Sigma^{-1}\frac{\partial^2 \Sigma}{\partial \delta_j \delta_k}] \tag{10}$$

at the value $\delta = \hat{\delta}$ and where $\Sigma = \Sigma(\delta)$. Note that δ and $V(\delta)$ can be estimated using only the empirical covariance S and not the individual Y_i, $1 \leq i \leq n$ (since S is sufficient for δ).

In practice, however, it is often not correct to assume that ζ and ϵ are multivariate normal. Under these circumstances, the application of the ML procedure based on the normal density is no longer justified in general. One approach to this problem is that of Arminger and Schoenberg (1989) based on the *pseudo maximum likelihood* (PML) method of Gourieroux, Montfort and Trognon (1984). It can be shown that under mild conditions, the "normal theory" ML estimator $\hat{\delta}$ still is consistent for δ_0 even if multivariate normality does not hold. But in general $(-A_n)^{-1}$ is not a "good" estimator of the asymptotic covariance matrix $V(\delta_0)$. A *consistent* estimator, however, will be

$$\hat{V}_P = A_n^{-1} B_n A_n^{-1\prime} \quad, \tag{11}$$

where

$$B_n = \frac{1}{n} \sum_{i=1}^{n} q(Y_i; \delta) q(Y_i; \delta)' \quad \text{at} \quad \delta = \hat{\delta} \tag{12}$$

and

$$q(y; \delta_j) = -\text{tr}[\Sigma^{-1}(yy' - \Sigma)\Sigma^{-1} \frac{\partial \Sigma}{\partial \delta_j}] \quad . \tag{13}$$

Note that in this case, the individual Y_i are needed for the estimate of $V(\delta_0)$.

As a special case, the following situation should be mentioned. Assume that in the model given above the disturbance terms ζ have an unrestricted covariance matrix and the components of ϵ are independent, i.e., Θ is diagonal. Then the ML method of estimating Λ, B, Ψ and Θ not only yields consistent estimators of the parameters, but also consistent estimators of the covariance matrices of the structural parameters Λ and B - even if the errors are not multivariate normal (Anderson and Amemiya 1988). Hence, in this case, the construction of asymptotic confidence intervals for components of the structural parameters using the ML method will be correct, even if the distributions of ζ and ϵ are misspecified.

3 Bootstrap estimates

For many problems, the bootstrap technique of estimating the distribution of an estimation error $(\hat{\delta} - \delta_0)$ turned out to be a valuable tool, especially for small sample sizes. Consider the "estimation error", i.e. the parametric function

$$R(Y, F) = \hat{\delta} - \delta(F) \quad . \tag{14}$$

We are looking for a "reasonable" estimate of its distribution

$$D_n(R|F) = D_F(R(Y_1, \ldots, Y_n; F)) \quad, \tag{15}$$

where F denotes the distribution of Y_i. The Bootstrap estimate is

$$\hat{D}_n(R|F) = D(R|\hat{F}_n) \quad, \tag{16}$$

where $\hat{F}_n$ is a suitable estimator of F. In the "classical" situation, $\hat{F}_n$ is the empirical distribution function $\hat{F}_n^{(Y)}$, say. Unfortunately, with probability one, $\hat{F}_n^{(Y)}$ does not possess the mean and covariance structure given by the model restrictions on F. Hence the bootstrap estimate $\hat{\delta}^\star$ of $\delta(\hat{F}_n^{(Y)})$ would not make sense. The problem is to find a "good" estimate of F that is compatible with the mean and covariance structure of the model assumptions. Now consider the random vectors

$$Z_i = (\Sigma(\hat{\delta}))^{1/2} S^{-1/2} (Y_i - \bar{Y}) \quad , \quad 1 \le i \le n \quad . \tag{17}$$

Under F, the distributions of Y_i and Z_i are very close - but the distribution of of Z_i is a member of the model family. Hence, we propose the Bootstrap estimate

$$\hat{D}_n(R|F) = D(R|\hat{F}_n^{(Z)}) \quad . \tag{18}$$

In the usual bootstrap technique, $\hat{D}_n$ is estimated using a Monte Carlo (MC) experiment that resamples from $Z_1, \ldots, Z_n$. Note that this approach is different to that of Boomsma (1986). But similarly, in the present approach some problems occur: First, the maximization algorithm used for estimation might stop at a local, not a global maximum. Hence bootstrapping does not "check" the behavior of the MLE but of the technical estimation procedure that might be different. Furthermore, in general there is a positive probability that the MLE does not exist at all and that the iteration algorithm does not converge. In bootstrap calculations, it is very probable that for at least some MC loops $\hat{\delta}^*$ does not exist.

One has to decide how to treat these cases: e.g. one way is to interrupt the iteration process of the ML estimation procedure (and hence consider only a modification of the MLE); another way might just skip all MC loops where the MLE fails to exist (and hence estimate a conditional distribution of the MLE given that it exists).

4 A Monte Carlo study

In this paper we do not consider mathematical properties of the bootstrap technique introduced in section 3. Usually these properties are of asymptotic nature; but then still we know little about the behavior for finite samples. Hence we want to compare the different ways of estimating the variance of the ML estimation of a parameter and check their biases and their ranges.

We consider the following model that is typical for many models in social science research:

$$B = \begin{pmatrix} 0 & 0.4 & 0 \\ 0 & 0 & 0.5 \\ 0 & 0 & 0 \end{pmatrix} \quad ,$$

$$\psi = \begin{pmatrix} 1.5 & 0 & 0 \\ 1.0 & 1.5 & 0 \\ 0 & 0 & 2.0 \end{pmatrix} \quad ,$$

$$\Lambda = \begin{pmatrix} 1.0 & 0 & 0 \\ 0.9 & 0 & 0 \\ 0.9 & 0 & 0 \\ 0 & 1.0 & 0 \\ 0 & 0.8 & 0 \\ 0 & 0.8 & 0 \\ 0 & 0 & 1.0 \\ 0 & 0 & 0.7 \\ 0 & 0 & 0.7 \end{pmatrix} \quad ,$$

$$\Theta = 0.6 \cdot I \quad .$$

Since the loadings λ_{11}, λ_{42} and λ_{73} have been set to 1.0, Y_1, Y_4 and Y_7 are reference indicators for η to fix the scale for the latent variables. The model corresponds to a path analytic model

$$\eta_3 \longrightarrow \eta_2 \longrightarrow \eta_1 \quad . \tag{19}$$

The errors of the dependent variables η_2 and η_1 are allowed to be correlated, ψ_{31} and ψ_{32} are 0 since the error terms for η_1 and η_2 must be uncorrelated with the regressor η_3. Each of the latent η–variables is measured by three indicators in a simple structural model. The error terms in the measurement model are independent and yield a diagonal matrix Θ. Hence, even in this simple structural model, there are $2 + 6 = 8$ structural and $4 + 9 = 13$ covariance parameters to be estimated.

Random vectors Y_i were generated in the MC study as follows. Generate 12 independent, identically distributed, standardized random variables $U_{i1}, \ldots, U_{i12}$ and let

$$\left[\begin{array}{c} \zeta \\ \epsilon \end{array} \right] = \left[\begin{array}{c} \Psi^{1/2} \\ \Theta^{1/2} \end{array} \right] \cdot U_i \quad . \tag{20}$$

As distributions of U_{ik}, we considered (1) the normal, (2) χ_1^2, and (3) the exponential distribution, each of them standardized by an affine transformation to a distribution with mean 0 and variance 1.

As sample sizes we considered: (a) $n = 100$, (b) $n = 200$, and (c) $n = 500$. The Monte Carlo study of the behavior of the MLE as well as the comparison of the ML and the PML estimation of $V(\hat{\delta})$ was done with the program LINCS (Schoenberg 1990) within GAUSS, which uses the Broyden-Fletcher-Goldfarb-Shannon (BFGS) algorithm to maximize the log-likelihood function. After 200 BFGS iteration steps, convergence was forced, i.e. we used the "modified" MLE described in section 3.

Evidently a long time is needed to do a MC study on the behavior of a bootstrap estimator, especially in our situation. We took $N_{BOOT} = 100$, $N_{MC} = 500$ and considered only the cases (1a), (1c), (2b), and (3b) (other combinations will be presented in a later paper). Each of these MC studies needed nearly four days of computing time on an 25 Mhz/486 PC.

5 Monte Carlo results

For four parameters ($\lambda_{21} = 0.9$, $\beta_{12} = 0.4$, $\psi_{11} = 1.5$ and $\theta_{11} = 0.6$), the following quantities have been estimated through the MC studies:

1. The mean of the MLE's.

2. The standard deviation of the MLE's denoted by s_{MC}.

3. Mean and standard deviation of the MLE s_{ML} of s_{MC}.

4. Mean and standard deviation of the PML s_{PML} of s_{MC}.

5. Mean and standard deviation of the bootstrap estimate s_B of s_{MC}.

The estimates of the structural parameters λ and β are given in table 1 for the cases (1a), (1c), (2b) and (3b); those for the covariance parameters θ and ψ in table 2. The estimates s_{ML} are incorrect for ψ and θ, but they are very good for the structural parameters λ and β (which is to be expected, since the assumptions of Anderson and Amemiya (1989) are satisfied in the model of section 4). In

Table 1. MC results: structural parameters

Distribution	$\mathcal{N}$	$\mathcal{N}$	χ^2	$Exp.$	$\mathcal{N}$	$\mathcal{N}$	χ^2	$Exp.$
Sample size	100	500	200	200	100	500	200	200
Parameter		$\lambda = 0.9$				$\beta = 0.4$		
E	.900	.900	.903	.902	.449	.397	.385	.389
s_{ML}	.073	.031	.054	.051	.171	.094	.157	.150
$E\hat{s}_{ML}$	.071	.032	.051	.051	.197	.093	.156	.150
$E\hat{s}_{PML}$	.071	.031	.050	.050	.200	.093	.147	.146
$E\hat{s}_B$	.071	.031	.051	.051	.259	.094	.156	.159
				standard deviations of				
$\hat{s}_{ML}$	.010	.002	.009	.007	.046	.013	.049	.038
$\hat{s}_{PML}$	.012	.002	.013	.010	.052	.013	.046	.037
$\hat{s}_B$	.014	.002	.013	.010	.115	.016	.051	.052

Table 2: MC results: covariance parameters

Distribution	$\mathcal{N}$	$\mathcal{N}$	χ^2	$Exp.$	$\mathcal{N}$	$\mathcal{N}$	χ^2	$Exp.$
Sample size	100	500	200	200	100	500	200	200
Parameter		$\psi = 1.5$				$\theta = 0.6$		
E	1.413	1.521	1.574	1.532	.585	.595	.600	.593
s_{ML}	.370	.238	.587	.456	.137	.058	.164	.138
$E\hat{s}_{ML}$	.451	.227	.408	.378	.313	.059	.093	.092
$E\hat{s}_{PML}$	.450	.228	.531	.444	.130	.059	.161	.130
$E\hat{s}_B$	.939	.234	.610	.538	.126	.059	.157	.130
				standard deviations of				
$\hat{s}_{ML}$	.166	.056	.252	.173	.018	.003	.016	.013
$\hat{s}_{PML}$	.175	.058	.338	.207	.023	.005	.069	.040
$\hat{s}_B$	1.086	.066	.452	.460	.022	.006	.064	.043

all cases, the PML estimate is fairly close to the true value, the variance being slightly bigger than that of the MLE (which is trivial, too).

The bootstrap estimate seems to be worse than PML in all cases. Even if the mean estimate is good, in general the variance is bigger. Finally, in the $n = 100$ case and in the estimation of $s_{MC}(\psi)$, bootstrapping fails completely. There are several possible explanations for this strange behavior. First of all, note that (by using $N_{BOOT} = 100$ instead of infinity) we considered only an estimate of the bootstrap estimate itself, this estimate adds a further part to the variance. Second, the simulation study has shown that especially in the case $n = 100$, the BFGS iterations had to be stopped fairly often - implying a somewhat "strange" estimate of the parameters. Here, especially ψ_{11} turned out to be extremely sensitive. In these cases, the "forced convergence" principle leads to an incorrect application of the bootstrap concept. Hence the idea of estimating the "conditional MLE distribution" (given the iteration process stops before 200) might be a better approach.

6 Conclusion

Since this paper only contains a simulation study, no mathematical result can be presented. The study, however, shows that in the model considered, PML seems to be better than bootstrap, especially in small sample situations. The bootstrap behavior seems to be bad mainly if the MLE does not exist with a (relatively) large probability. Furthermore, the MC procedure in estimating the bootstrap estimate seems to add an error term that cannot be neglected in the present situation.

References

Anderson, T.W. and Amemiya, Y. (1988). The asymptotic normal distribution of estimators in factor analysis under general conditions. *Annals of Statistics* 16, 759 - 771.

Arminger, G. and Schoenberg, R. (1989). Pseudo maximum likelihood estimation and a test for misspecification in mean- and covariance structure models. *Psychometrika* 54, 409 - 425.

Boomsma, A. (1986). *On the use of Bootstrap and Jackknife in covariance structure analysis*. In: *Compstat 1986*, Heidelberg: Pysica-Verlag.

Gourieroux, C., Monfort, A. and Trognon, A. (1984). Pseudo maximum likelihood methods: Theory. *Econometrica* 52, 681 - 700.

Jöreskog, K.G. and Sörbom, D. (1988). *LISREL 7: A guide to the program and applications*. Chicago: SPSS Inc.

Schoenberg, R. (1990). *LINCS 2.0: Linear covariance structure analysis. User's guide*. Kent WA: Aptech Systems.

Bootstrap: Selected Topics (7)

ON THE BAYESIAN BOOTSTRAP

Raul Cano
Department of Statistics, Stockholm University
S-106 91 Stockholm, Sweden

SUMMARY

A method to compare the Bayesian bootstrap with a parametric analysis is derived. The method is applied to the mean of a Poisson distribution.

1. INTRODUCTION

Suppose that two Statisticians **A** and **B** have observed n   i.i.d. random variables $X = (X_1, \ldots, X_n)$ from an unknown distribution **F**. They are both interested in obtaining the posterior distribution of $\Psi(F)$, a functional depending of **F**.

Statistician **A** knows the form of the distribution and uses a parametrical method. On the other hand, Statistician **B** does not know anything about the underlying distribution and uses the Bayesian bootstrap method (Rubin, 1981).

In section 3, it is given a method to compare the approximate method used by Statistician **B** with the " optimal " method used by Statistician **A**. The method is applied to the mean functional $\Psi(F) = \int y\ F(dy)$, assuming the underlying distribution **F** to be a Poisson with parameter θ.

2. THE BAYESIAN BOOTSTRAP

Let α be a real finite measure. We write $F \sim \mathbb{D}_\alpha$ to mean that **F** is a random distribution function that is distributed as a Dirichlet process with parameter α (see Ferguson [1973]). As usual, we do not distinguish between a random variable and the value it takes. Then **F|X** will denote the random distribution function **F** given $X = x$ and $\Psi(F)|X$ the random variable $\Psi(F)$ given $X = x$.

The Bayesian bootstrap method (Rubin, 1981) consists of using the distribution of $\Psi(D_n)|X$ to approximate the distribution of $\Psi(F)|X$, where $D_n|X \sim \mathbb{D}_{\Sigma_i \delta_{X_i}}$, δ_x being the measure giving mass one to x. $D_n|X$ can also be described as a multivariate Dirichlet distribution concentrated on the observed sample.

It is easy to simulate the distribution of $\Psi(D_n)|X$ for any given functional Ψ using the Monte Carlo method (see Rubin, 1981).

The estimates of Statistician B are based on the distribution of $\Psi(D_n)|X$. From the following theorem it follows that his estimates are generalized Bayes rules in the sense of Ferguson (1967). Note that this does not mean that a decision rule based on $\Psi(D_n)|X$ is an admissible rule or a proper Bayes rule.

THEOREM 2.1   Let γ be a finite, absolutely continuous measure on $(\mathbb{R}, \mathcal{B})$. Define for $r \in \mathbb{N}$ and $B \in \mathcal{B}$, the measure β_r by $\beta_r(B) = \dfrac{\gamma\{\, z|\ rz \in B\ \}}{r}$,

if $F_r \sim \mathbb{D}_{\beta_r}$ then $F_r|X$ converges weakly to $D_n|X$ as $r \to \infty$.

PROOF. Ferguson (1973) showed that $F_r|X \sim \mathbb{D}_{\beta_r + \Sigma_i \delta_{X_i}}$. Let $\alpha_r = \beta_r + \Sigma_i \delta_{X_i}$

and $\alpha_0 = \Sigma_i \delta_{X_i}$, then $\underset{B \in \mathcal{B}}{\sup} \; | \; \alpha_r(B) - \alpha_0(B) \; | = \dfrac{\gamma(\mathbb{R})}{r} \to 0$ as $r \to \infty$

and by Theorem 3.2 of Sethuraman and Tiwari (1982) we obtain that $F_r|X$ converges weakly to $D_n|X$.

3.   A NUMERICAL STUDY OF THE BAYESIAN BOOTSTRAP ON THE MEAN FUNCTIONAL OF THE POISSON DISTRIBUTION.

In this section, it is assumed that the underlying distribution F is Poisson with parameter θ and that Statisticians A and B are interested in the mean functional $\Psi(F) = \int y\, F(dy) = \theta$.

Statistician B obtains the distribution of $\Psi(D_n)|X$ by means of the Monte Carlo method. In this case $\Psi(D_n)|X = X_1 g_1 + \ldots + X_n g_n$ where $(g_1, \ldots, g_n)$ follows a Dirichlet distribution with parameter $(1, \ldots, 1)$.

On the other hand, and in order to make the analysis comparable, Statistician A uses a non-informative prior for θ such that $E(\Psi(F)|X) = E[\Psi(D_n)|X]$. Moreover, for mathematical convenience he bases his analysis on the conjugate prior for θ. Then, taking $\theta \sim \text{gamma}(0, 0)$ he obtains that $\Psi(F)|X \sim \text{gamma}(\Sigma\, X_i, n)$.

The first two central moments of the posterior dist. for Statistician B are

$$E[\Psi(D_n)|X] = \Sigma\, X_i / n = \bar{X} \quad \text{and} \quad \text{VAR}[\Psi(D_n)|X] = \Sigma\, (X_i - \bar{X})^2 / n(n+1)$$

and, for Statistician A, are

$$E(\Psi(F)|X) = \Sigma\, X_i / n \quad \text{and} \quad \text{VAR}(\Psi(F)|X) = \Sigma\, X_i / n^2.$$

By requirement, the first moments agree, which indicates that the non-informative priors of the two Statisticians contain roughly the same information. That the second moments disagree , reflects the fact that Statistician A knows the form of the underlying distribution. Moreover, we have that

$$E (VAR[\Psi(D_n)|X]) = \lambda(n-1) / n(n+1) \qquad \text{and}$$

$$E (VAR(\Psi(F)|X)) = \lambda / n.$$

Thus, the posterior distribution of Statistician B is expected to be more concentrated than the distribution of Statistician A. One would wish for the opposite to be true.

For the purpose of illustration, we simulated 20 values from a Poisson random variable with parameter $\lambda = 1$. The following values were obtained 3,2,0,1,1,1,0,1, 1,1,0,1,1,0,1,0,1,1,0,0.

Figure 1 shows a comparison between the histogram of $\Psi(D_n)|X$ (based on 10,000 replicates) and the density of $\Psi(F)|X$, which in this case is a gamma density with parameters a = 16 and b = 20. We can see that the histogram is skewed to the right in the same way as the density under the parametric Poisson-gamma model.

Since the histogram of $\Psi(D_n)|X$ approximates the density of $\Psi(F)|X = \theta$, the histogram gives for Statistician B an approximated $(1-\alpha)$ HPD (highest posterior density) interval for θ.

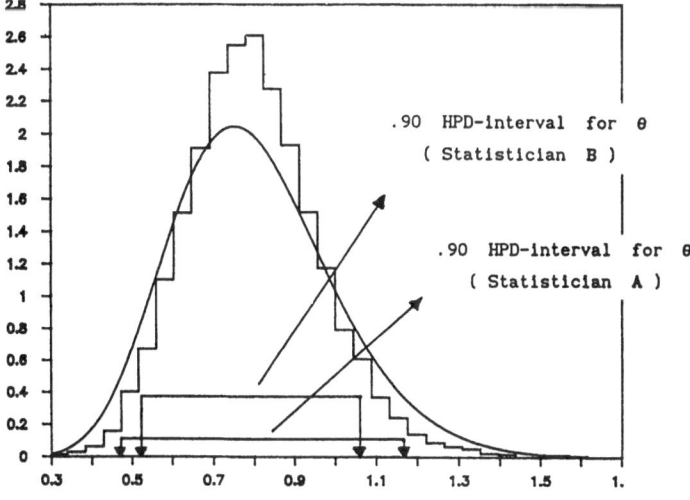

Figure 1. Histogram of $\Psi(D_n)|X$ and density of $\Psi(F)|X$.

Taking $\alpha = .10$, he obtains the approximated .90 HPD-interval [.52 , 1.06]. That is, if he just consider the information supplied by the histogram, he can say that with probability .90 the true value of θ lies in [.52 , 1.06], which is a fairly good interval when compared with the exact .90 HPD-interval [.474 , 1.117], which is obtained by Statistician **A**.

In order to see how the quantiles of the Bayesian bootstrap performs with respect to the " optimal " Bayesian parametrical analysis, we have run a Monte Carlo experiment by drawing 1,000 random samples (trials) of fixed size n from the Poisson model with $\theta = 1$ and then bootstrapping the mean functional for each trial with 10,000 replicates.

Next, for each trial i, we have obtained (as Statistician **B**) 99 percentiles $\varepsilon_{\alpha, i}$ from the corresponding histogram, where $\alpha = .01, .02, \ldots , .98, .99$, i = 1, ... ,1000. Further, we have computed $P(\Psi(F) \leq \varepsilon_{\alpha, i} | X) = P_{\alpha, i}$ under the parametric Poisson-gamma model (i.e. $\Psi(F)|X \sim \text{gamma}(\Sigma X_i, n)$). We have also computed $\bar{P}_\alpha = \sum_{i=1}^{1000} P_{\alpha, i} / 1000$, where α represents the intended probability and $\bar{P}_\alpha$ the average of the observed posterior probability. If the Bootstrap method gave correct probabilities, $\bar{P}_\alpha$ should be equal to α.

We have carried out the above procedure for the following different sample sizes: n = 20, 50, 100. The results are summarized in Fig.2 which displays the graphs of α against the bias $\bar{P}_\alpha - \alpha$ and the standard deviation of $\bar{P}_\alpha$ respectively. For other values of θ the results are quite similar to those in Fig.2.

4. CONCLUSIONS

The performance of the Bayesian bootstrap is given in Fig.2. From these numerical results we can say that for moderate sample sizes, in average, the quantiles of the Bayesian bootstrap overestimate the intended probability α for $\alpha < .55$ and they underestimate α for $\alpha > .65$. Our results can be used to improve the precision of the quantiles of the Bayesian bootstrap, but the details are not included here.

REFERENCES

Ferguson, Thomas S. (1967). Mathematical Statistics. Academic Press 1967.
Ferguson, Thomas S. (1973). A Bayesian analysis of some nonparametric problems. Ann. Statist.1 ^09-230.
Rubin, Donald B. (1981 The Bayesian bootstrap. Ann. Statist.9 130-134.
Sethuraman, J. and Tiwari, R. (1982). Convergence of Dirichlet measures and the interpretation of their parameter. In : Statistical decision theory and related topics III, Vol.2, 305-315, Gupta S.S. and Berger J.O. (eds.). New York : Academic Press 1982.

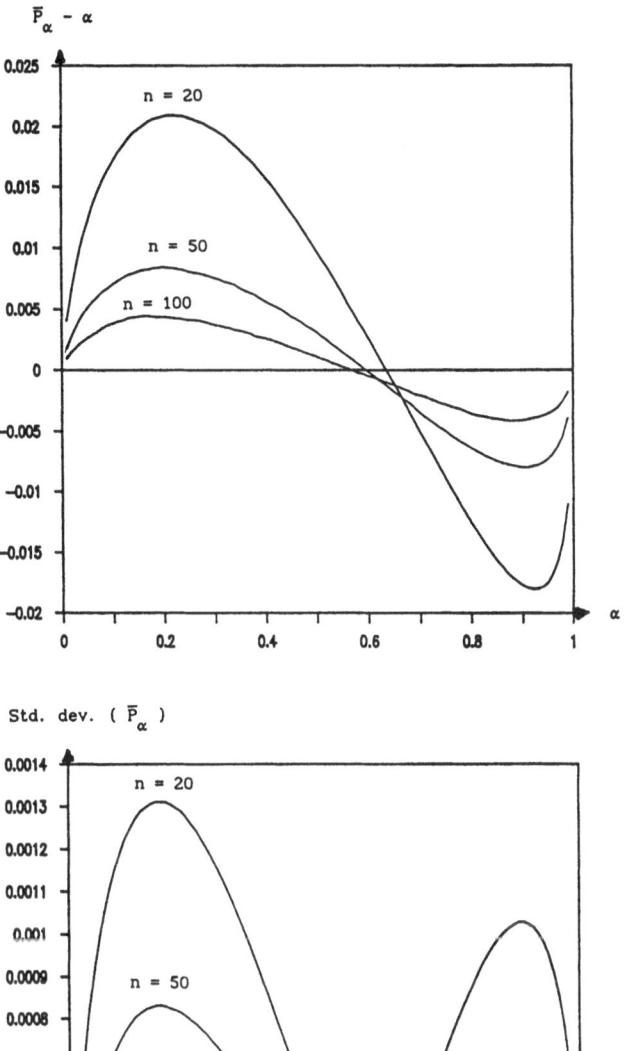

Figure 2. Graphs of α against the bias $\bar{P}_\alpha - \alpha$ and the standard deviation of $\bar{P}_\alpha$ for 99 values of α, α = .01, .02, ..., .98, .99. The resultant points were connected by straight lines.

BOOTSTRAPPING THE SAMPLE QUANTILE: A SURVEY

Michael Falk
Katholische Universität Eichstätt
Mathematisch-Geographische Fakultät
Ostenstr. 26–28, D–8078 Eichstätt
FALK@URZ.KU-EICHSTAETT.DBP.DE

1. The Bootstrap Estimate. Let $X_1, \ldots, X_n$ be independent and identically distributed ($\equiv$ iid) random variables ($\equiv$ rvs) with common distribution function ($\equiv$ df) F and let $T(F)$ be an unknown parameter of interest. The natural nonparametric estimator of $T(F)$ is $T(F_n)$, where $F_n(t) := n^{-1} \sum_{i=1}^{n} 1_{(-\infty,t]}(X_i)$, $t \in \mathbb{R}$, denotes the empirical df pertaining to the sample $X_1, \ldots, X_n$. Our target function is now the df of the estimator $T(F_n)$, centered at the unknown parameter $T(F)$, i.e.

$$G_n(x) := P_F\{n^{1/2}(T(F_n) - T(F)) \le x\}, \quad x \in \mathbb{R}.$$

In a great many of cases, the factor $n^{1/2}$ is the corrected standardization constant to ensure the nondegenerate limit distribution of $n^{1/2}(T(F_n) - T(F))$ as n increases.

Note that $G_n(\cdot)$ is itself a functional of the underlying df F, i.e. $G_n(\cdot) = \mathbf{T}(F)$, and therefore, the natural nonparametric estimate of G_n is $\mathbf{T}(F_n) =: G_n^*(\cdot)$, with

$$G_n^*(x) = P_{F_n}\{n^{1/2}(T(F_n^*) - T(F_n)) \le x\}, x \in \mathbb{R},$$

where F_n^* is the empirical df pertaining to a sample $X_1^*, \ldots, X_n^*$ of iid rvs with common df F_n. Since Efron (1979) the estimator G_n^* is known as the *bootstrap estimator* of G_n. The bootstrap estimate therefore turns out to be the classical approach of nonparametric statistics, where the underlying but unknown df is replaced by the empirical df pertaining to the data. The new but crucial point is the type of highdimensional functional $\mathbf{T}(F)$ with values in the space of dfs which we are going to estimate by $\mathbf{T}(F_n)$. In addition, a Monte-Carlo simulation is usually inherent in the computation of the estimator $\mathbf{T}(F_n)$.

This particular problem within a classical statistical framework entails a list of fascinating consequences and we will present several of these for the particular functional

$$T(F) = F^{-1}(q) = \inf\{t \in \mathbb{R} : F(t) \ge q\},$$

where $q \in (0,1)$ is fixed. Then $T(F_n)$ is the sample quantile $F_n^{-1}(q)$ and we know that the pertaining bootstrap estimate

$$G_n^*(x) = P_{F_n}\{n^{1/2}(F_n^{*-1}(q) - F_n^{-1}(q)) \le x\}, \quad x \in \mathbb{R},$$

has a poor rate of convergence which is captured in the following two theorems. By $\to_D$ we denote convergence in distribution.

1.1 Theorem (Singh (1981)). *If F has a bounded second derivative in a neighborhood of $F^{-1}(q)$ and $F'(F^{-1}(q)) > 0$, then a.s.*

$$\limsup_{n \to \infty} n^{1/4} (\log \log n)^{-1/2} \sup_{x \in \mathbb{R}} \mid G_n^*(x) - G_n(x) \mid = K_{q,F},$$

where the constant $K_{q,F}$ depends only upon q and F.

1.2 Theorem (Falk and Reiss (1989 a)). *Suppose that F is differentiable near $F^{-1}(q)$ such that $f = F'$ is Hölder–continuous of order $\delta > 1/2$ and $f(F^{-1}(q)) > 0$. Then the bootstrap process $(G_n^*(x) - G_n(x))_{x \in \mathbb{R}}$, as a random element in the space $D_{\mathbb{R}}$ of those functions on $\mathbb{R}$ which are right-hand continuous and have left–hand limits, suitably standardized converges weakly to the Brownian motion B on $\mathbb{R}$:*

$$n^{1/4} \left(\psi(x) \left(G_n^*(x) - G_n(x)\right)\right)_{x \in \mathbf{R}} \to_D \left(B(x)\right)_{x \in \mathbf{R}} = \left((B_1(-x))_{t \leq 0}, (B_2(x))_{t > 0}\right),$$

where B_1, B_2 are independent Brownian motions on $[0, \infty)$ and the weight function ψ is given by $\psi(x) := (q(1-q))^{1/2} f(F^{-1}(q))^{-1/2}/\varphi\left((q(1-q))^{-1/2} f(F^{-1}(q))x\right)$ for $x \in \mathbf{R}$, φ denoting the standard normal density.

We remark that a slight modification of the original proof in Falk and Reiss (1989 a) entails that continuity of F is actually needed only in a neighborhood of $F^{-1}(q)$. Weak convergence of the maximum error $n^{1/4} \sup_{x \in \mathbf{R}} |G_n^*(x) - G_n(x)|$ to the respective functional $\sup_{x \in \mathbf{R}} |B(x)/\psi(x)|$ has been established by Falk (1990a). Note that this result is *not* an immediate consequence of Theorem 1.2 and the continuous mapping theorem.

The preceding results show that the bootstrap estimate has a poor rate of convergence $O(n^{-1/4})$, roughly, in case of the quantile, whereas the bootstrap estimate pertaining to the mean has rate of convergence $O(n^{-1/2})$, or $O(n^{-1})$, roughly, for the studentized mean (Singh (1981)). Recall that the sample quantile itself has accuracy of order $O(n^{-1/2})$ and therefore the accuracy of the bootstrap need not coincide in general with the accuracy of the functional estimate behind.

The reason behind this completely different behavior of the bootstrap estimate for the quantile is that the bootstrap is actually a naive density estimate in this case. Recall that for any df H the random variable $H^{-1}(U)$ has df H if U is uniformly on (0,1) distributed. This representation together with the equivalence $H^{-1}(p) \leq x \Leftrightarrow p \leq H(x)$ and the Berry-Essen theorem for the sample quantile (cf. Theorem 4.2.1 in Reiss (1989)) implies

$$G_n^*(x) = P_{F_n}\{n^{1/2}(F_n^{*-1}(q) - F_n^{-1}(q)) \leq x\} = P_n\{F_n^{-1}(\overline{F}_n^{-1}(q)) \leq F_n^{-1}(q) + xn^{-1/2}\}$$

$$= P_n\{\overline{F}_n^{-1}(q) \leq F_n(F_n^{-1}(q) + xn^{-1/2})\} = P_n\{n^{1/2}(\overline{F}_n^{-1}(q) - q) \leq \frac{F_n(F_n^{-1}(q) + xn^{-1/2}) - q}{n^{-1/2}}\}$$

$$= \Phi\left(\frac{F_n(F_n^{-1}(q) + xn^{-1/2}) - q}{n^{-1/2}(q(1-q))^{1/2}}\right) + O(n^{-1/2}),$$

where $\overline{F}_n$ denotes the empirical df pertaining to a sample of n independent and uniformly on (0,1) distributed rvs, and Φ is the standard normal df. We add the index n to the above probabilities to indicate their dependence on the outcome of F_n. Hence,

$$G_n^*(x) \sim \Phi\left(\frac{F_n(F_n^{-1}(q) + xn^{-1/2}) - F_n(F_n^{-1}(q))}{xn^{-1/2}} \frac{x}{(q(1-q))^{1/2}}\right), \tag{1.3}$$

where

$$g_n(x) := \frac{F_n(F_n^{-1}(q) + xn^{-1/2}) - F_n(F_n^{-1}(q))}{xn^{-1/2}}$$

is a histogram type estimator of the density quantile function $f(F^{-1}(q))$, $f = F'$.

By repeating the above arguments we obtain the corresponding representation for the target function $G_n(x)$:

$$G_n(x) = \Phi\left(\frac{F(F^{-1}(q) + xn^{-1/2}) - q}{n^{-1/2}(q(1-q))^{1/2}}\right) + O(n^{-1/2}) \sim \Phi\left(f(F^{-1}(q)) \frac{x}{(q(1-q))^{1/2}}\right),$$

In case of $T(F) = F^{-1}(q)$, the bootstrap estimate G_n^* is therefore closely related to the density estimate g_n:

$$G_n^*(x) - G_n(x)$$

$$\sim \Phi\left(g_n(x) \frac{x}{(q(1-q))^{1/2}}\right) - \Phi\left(f(F^{-1}(q)) \frac{x}{(q(1-q))^{1/2}}\right)$$

$$\sim \varphi \left(f(F^{-1}(q)) \frac{x}{(q(1-q))^{1/2}} \right) \frac{x}{(q(1-q))^{1/2}} \left(g_n(x) - f(F^{-1}(q)) \right)$$

by Taylor expansion. This explains several of its properties (cf. section 6.4 of the book by Reiss (1989) for further details).

This observation suggests the idea that the bootstrap approach might be improved by involving a refined density estimate. This leads to the idea of a *smoothed bootstrap*. Put for $t \in \mathbb{R}$

$$\hat{F}_n(t) := \int K((t-x)/\alpha_n) \, F_n(dx) = n^{-1} \sum_{i=1}^{n} K((t-X_i)/\alpha_n),$$

where the kernel function K is typically a df and the positive bandwidth α_n tends to zero as n increases. The kernel estimate $\hat{F}_n$ of the underlying df F is a smoothed version of the empirical df which is superior to F_n under suitable conditions (cf. Reiss (1981), Falk (1983), Jones (1990)).

If we replace now the resampling df F_n by $\hat{F}_n$ we obtain the smoothed bootstrap estimate $\mathbf{T}(\hat{F}_n) =: \hat{G}_n^*$ of $G_n = \mathbf{T}(F)$:

$$\hat{G}_n^*(x) = P_{\hat{F}_n}\{n^{1/2}(T(F_n^*) - T(\hat{F}_n)) \le x\}, \quad x \in \mathbb{R}.$$

Again F_n^* is the empirical df pertaining to a sample $X_1^*, \ldots, X_n^*$ of iid rvs, but this time these are generated according to $\hat{F}_n$.

This particular version of a smoothed empirical df allows us to generate $X_1^*, \ldots, X_n^*$ in a simple way: If $Y_1^*, \ldots, Y_n^*$ are independent rvs with common df F_n and $V_1, \ldots, V_n$ are independent rvs with common df K and if these two samples are independent, then we may take $X_i^* := Y_i^* + \alpha_n V_i$, $i = 1, \ldots, n$. Therefore, the Monte-Carlo simulation behind the smoothed bootstrap does not require substantially more computing capacity than the ordinary bootstrap.

By repeating the arguments which led to formula (1.3) we obtain in the case $T(F) = F^{-1}(q)$, if the kernel K is differentiable with derivative k

$$\hat{G}_n^*(x) \sim \Phi \left(\frac{\hat{F}_n(\hat{F}_n^{-1}(q) + x n^{-1/2}) - \hat{F}_n(\hat{F}_n^{-1}(q))}{n^{-1/2}(q(1-q))^{1/2}} \right) \sim \Phi \left(\hat{f}_n(F^{-1}(q)) \frac{x}{(q(1-q))^{1/2}} \right)$$

by Taylor–expansion, where

$$\hat{f}_n(t) := \hat{F}_n'(t) = \int \alpha_n^{-1} k((t-x)/\alpha_n) \, F_n(dx) = (n\alpha_n)^{-1} \sum_{i=1}^{n} k((t-X_i)/\alpha_n)$$

is the kernel density estimate of the density f of F which we assume to exist. Under suitable smoothness conditions on F it is therefore clear that the smoothed bootstrap $\hat{G}_n^*$ outperforms the ordinary one. This is captured in the following result which is Corollary 2.7 (ii) in Falk and Reiss (1989 b).

1.4 Theorem. *Suppose that F is thrice differentiable near $F^{-1}(q)$ with $f(F^{-1}(q)) > 0, f = F'$. Let $K : \mathbb{R} \to [0,1]$ have support $[-1,1]$, be three times differentiable with bounded third derivative and satisfy $\int k(x)dx = 1, \int x k(x)dx = 0$, where $k = K'$. Then, if $n\alpha_n^3 \to_{n\to\infty} \infty$ and $n\alpha_n^5 \log^2(n) \to_{n\to\infty} 0$ we have*

$$(n\alpha_n)^{1/2} \sup_{x \in \mathbb{R}} | \hat{G}_n^*(x) - G_n(x) | \to_D (f(F^{-1}(q)) 2\pi e)^{-1/2} | X |,$$

where X is a standard normal random variable.

Notice that the accuracy of the smoothed bootstrap approximation is now roughly $O(n^{-2/5})$ for an appropriate choice of α_n. An analogous result for the bootstrap variance estimate for the sample quantile has been established by Hall et al. (1989).

A different smoothed bootstrap estimate can be based on a smoothed version of the sample quantile function ($\equiv$ qf) F_n^{-1}. Put

$$\tilde{F}_n^{-1}(p) := \int_0^1 F_n^{-1}(x)\alpha_n^{-1}k((p-x)/\alpha_n)\,dx = \alpha_n^{-1}\sum_{i=1}^n \left(\int_{(i-1)/n}^{i/n} k((p-x)/\alpha_n)\,dx\right) X_{i:n},$$

where $X_{1:n} \leq \ldots \leq X_{n:n}$ are the order statistics pertaining to the sample $X_1,\ldots,X_n$, the kernel $k : \mathbb{R} \to [0,\infty)$ with support $[-1,1]$ satisfies $\int k(x)dx = 1$ and $\alpha_n \to_{n\to\infty} 0$. $\tilde{F}_n^{-1}(p)$ is monotone increasing on $[\alpha_n, 1 - \alpha_n]$ and a modification on $(0,\alpha_n) \cup (\alpha_n,1)$ makes it a qf. Under suitable conditions, $\tilde{F}_n^{-1}(p)$ is superior to the empirical qf (Falk (1984), chapter 8 of the book by Reiss (1989)).

As the resampling df in the bootstrap approach we now take the pertaining df $\tilde{F}_n$, yielding the smoothed bootstrap estimate

$$\tilde{G}_n^*(B) := P_{\tilde{F}_n}\left\{n^{1/2}(T(F_n^*) - T(\tilde{F}_n)) \in B\right\}, \quad B \in \mathbb{B},$$

of

$$G_n(B) := P\{n^{1/2}(T(F_n) - T(F)) \in B\}, \quad B \in \mathbb{B},$$

where $\mathbb{B}$ denotes the Borel-σ-algebra of $\mathbb{R}$. Notice that we can generate X_i^* by choosing $X_i^* = \tilde{F}_n^{-1}(U_i)$, $U_1, U_2,\ldots$ being independent and uniformly on $(0,1)$ distributed rvs.

For $T(F) = F^{-1}(q)$ we can now prove consistency even uniformly over all Borel sets of the smoothed bootstrap estimate $\tilde{G}_n^*$ (cf. Theorem 2.1 in Falk and Reiss (1989 b)).

1.5 Theorem. *Suppose that the kernel k is positive with support $[0,1]$, $\int k(x)\,dx = 1$ and that it has a bounded second derivative. Let α_n satisfy $n\alpha_n^3 \to_{n\to\infty} 0, n\alpha_n^2/\log(n) \to_{n\to\infty} \infty$. Then, if F^{-1} has a bounded second derivative near q we have uniformly over $x \geq 0$*

$$P\{(n\alpha_n/\int k^2(y)\,dy)^{1/2}(\pi e/2)^{1/2}\sup_{B\in\mathbb{B}}\mid \tilde{G}_n^*(B) - G_n(B)\mid\leq x\} \to_{n\to\infty} 2\Phi(x) - 1.$$

The preceding result holds uniformly for those df in the class $\mathcal{F}(\varepsilon, D_1, D_2, D_3) := \{F$ df $: F^{-1}$ has a second derivative on the interval $I(\varepsilon) := (q - \varepsilon, q + \varepsilon)$ with $D_1 \leq \inf_{p\in I(\varepsilon)}(F^{-1})'(p) \leq \sup_{p\in I(\varepsilon)}(F^{-1})'(p) \leq D_2$, $\sup_{p\in I(\varepsilon)}|(F^{-1})^{(2)}(p)| \leq D_3\}$, where D_1, D_2, D_3 are given positive constants.

Notice that in Theorem 1.5 the normalizing constants of the bootstrap error do not depend upon the underlying df, thus entailing immediately the construction of asymptotically distribution free confidence intervals for the maximum bootstrap error. The rate of convergence of this smoothed bootstrap estimate is now roughly $O(n^{-1/3})$. Finally, this approach enables us to establish consistency of the bootstrap estimate of the df of the maximum bootstrap error itself in this case: Put

$$H_n(F,x) := P_F\{\sup_{B\in\mathbb{B}}\mid \tilde{G}_n^*(B) - G_n(B)\mid\leq x\}, x \in \mathbb{R}.$$

Then a smoothed bootstrap estimate of $H_n(F,x)$ is $H_n(\tilde{F}_n, x)$ and we have the following result (Theorem 3.1 of Falk and Reiss (1989 b)).

1.6 Theorem. *If α_n, k and F satisfy the conditions of Theorem 1.5 we have with probability one*

$$\sup_{x\geq 0}\mid H_n(\tilde{F}_n,x) - H_n(F,x)\mid\to_{n\to\infty} 0.$$

One should however not neglect that the price one has to pay for the advantages of using $\tilde{F}_n$ is the fact that the computation of the pertaining smoothed bootstrap needs essentially more computer time than the one which uses $\hat{F}_n$.

2. Bootstrap confidence intervals.

Let again $G_n^*(x) = P_{F_n}\{n^{1/2}(T(F_n^*) - T(F_n)) \leq x\}$ be the ordinary bootstrap estimate of $G_n(x) = P_F\{n^{1/2}(T(F_n) - T(F)) \leq x\}$. If we knew G_n exactly, then the interval

$$I_n := (T(F_n) - G_n^{-1}(1-\alpha)/n^{1/2}, T(F_n) - G_n^{-1}(\alpha)/n^{1/2}]$$

would define a confidence interval ($\equiv$ ci) of exact level $1 - 2\alpha$ for the parameter $T(F)$ (if G_n is continuous):

$$P\{T(F) \in I_n\} = 1 - 2\alpha.$$

Clearly, G_n is usually unknown as $T(F)$ is, and so we replace G_n^{-1} in the definition of the interval I_n by the bootstrap estimate G_n^{*-1} thus obtaining the bootstrap ci

$$I_n^* := (T(F_n) - G_n^{*-1}(1-\alpha)/n^{1/2}, T(F_n) - G_n^{*-1}(\alpha)/n^{1/2}].$$

If $n^{1/2}(T(F_n) - T(F))$ has a continuous, strictly increasing limiting distribution and the bootstrap estimate G_n^* is (weakly) consistent, then I_n^* is a ci of *asymptotic* level $1 - 2\alpha$:

$$P\{T(F) \in I_n^*\} \to_{n\to\infty} 1 - 2\alpha.$$

It turns out that the rate of convergence of the bootstrap estimate itself and of the level error of I_n^* may differ considerably. We know from the results of the first section that the bootstrap estimate G_n^* pertaining to $T(F) = F^{-1}(q)$ has accuracy of order $O(n^{-1/4})$. One might guess therefore that this is also the level error of the corresponding bootstrap ci. It turns out, however, that I_n^* has actually a level error of surprisingly low order $O(n^{-1/2})$! This result is *not* caused by the fact that I_n^* is a twosided interval, as can readily be seen if we formulate this result in terms of the concept of *prepivoting*:

From the convergence

$$P\{\alpha \le G_n^* \left(n^{1/2}(T(F_n) - T(F))\right) < \beta\} = P\{G_n^{*-1}(\alpha) \le n^{1/2}(T(F_n) - T(F)) < G_n^{*-1}(\beta)\}$$

$$\to_{n\to\infty} \beta - \alpha$$

for $0 < \alpha < \beta < 1$ we obtain that the limit distribution of $G_n^* \left(n^{1/2}(T(F_n) - T(F))\right)$ is the uniform distribution on (0,1). This limit theorem in turn suggests the confidence interval

$$C_n := \{\vartheta \in I\!R : \alpha \le G_n^* \left(n^{1/2}(T(F_n) - \vartheta)\right) < 1 - \alpha\}$$

which coincides with I_n^*. Note that the rate at which $G_n^* \left(n^{1/2}(T(F_n) - T(F))\right)$ becomes uniformly on (0,1) distributed coincides with the level error of I_n^*.

This elementary idea of transforming $n^{1/2}(T(F_n) - T(F))$ by its *estimated* df is due to Beran (1987), who calls it *prepivoting* for obvious reasons. Recall that a rv which is transformed by its continuous df is exactly uniformly on (0,1) distributed. Prepivoting is in a certain sense equivalent to studentizing: If $n^{1/2}(T(F_n) - T(F))$ is asymptotically $N(0, \sigma^2)$-distributed with unknown $\sigma^2 > 0$, then $n^{1/2}(T(F_n) - T(F))/\hat\sigma_n$ is asymptotically standard normal for any consistent estimator sequence $\hat\sigma_n$ of σ, i.e. studentization, like prepivoting, provides an asymptotically distribution free statistical procedure.

For $T(F) = F^{-1}(q)$ prepivoting yields

$$G_n^*(n^{1/2}(T(F_n) - T(F))) = P_{F_n}\{n^{1/2}(F_n^{*-1}(q) - F_n^{-1}(q)) \le n^{1/2}(F_n^{-1}(q) - F^{-1}(q))\}$$

and we have the following result which is due to Falk and Kaufmann (1989).

2.1 Theorem. *Suppose that F is twice continuously differentiable near $F^{-1}(q)$ and that $f = F'$ is positive. Then we have for $x \in (0,1)$*

$$P\{G_n^*(n^{1/2}(F_n^{-1}(q) - F^{-1}(q))) < x\} = x + C_n(F, q, x)n^{-1/2} + o(n^{-1/2}),$$

where C_n is bounded in x and n. (A detailed description of C_n is given in Falk and Kaufmann (1989)).

The proof of the preceding result shows that that Theorem 2.1 holds uniformly for those df in the class $\mathcal{F}(\varepsilon, D_1, D_2) := \{F \text{ df} : F^{-1} \text{ is three times differentiable in } I(\varepsilon) = (q - \varepsilon, q + \varepsilon) \text{ with } \sup_{p \in I(\varepsilon)} | (F^{-1})^i(p) | \leq D_1, i = 1, 2, 3, (F^{-1})'(q) \geq D_2\}$, where D_1, D_2 are given positive constants.

The preceding result immediately entails that I_n^* as well as one–sided cis $(F_n^{-1}(q) - G_n^{*-1}(1 - 2\alpha)/n^{1/2}, \infty]$, $(-\infty, F_n^{-1}(q) - G_n^{*-1}(2\alpha)/n^{1/2}]$ for $F^{-1}(q)$ have level errors of precise order $O(n^{-1/2})$.

The prepivoting procedure can be iterated by nested bootstrap: the df of $G_n^*(n^{1/2}(T(F_n) - T(F)))$ can itself be estimated by the bootstrap approach; and if we denote the corresponding estimate by $H_n^*(x)$, then $H_n^* \left(G_n^*(n^{1/2}(T(F_n) - T(F))) \right)$ is the resulting (doubly prepivoted) quantity. This procedure can be iterated and the level errors of the pertaining cis may decrease, as Beran (1987) points out. The computing time for their derivation increases however exponentially fast.

It is quite surprising that the prepivoted sample quantile can even be made a pivot by a simple backward procedure; this provides on the other hand an explanation for the low level error of the ci I_n^*. Consider $G_n^*(n^{1/2}(T(F) - T(F_n)))$ for general $T(F)$ in place of $G_n^*(n^{1/2}(T(F_n) - T(F)))$. If $n^{1/2}(T(F_n) - T(F)))$ has a limiting continuous distribution which is symmetric to the origine, then $G_n^*(n^{1/2}(T(F) - T(F_n)))$ will approximately be uniformly on $(0,1)$ distributed as well. The resulting ci for $T(F)$ is then

$$I_{nb}^* := \{\vartheta : \alpha \leq G_n^*(n^{1/2}(\vartheta - T(F_n))) < 1 - \alpha\} \tag{2.2}$$

$$= \left[T(F_n) + G_n^{*-1}(\alpha)n^{-1/2}, T(F_n) + G_n^{*-1}(1 - \alpha)n^{-1/2} \right),$$

which arises from I_n^* by flipping its endpoints around $T(F_n)$, i.e. the endpoints of I_{nb}^* are *backward critical points* in the sense of Hall (1988); Efron (1982) calls this approach *percentile method*.

The intervals I_n^* and I_{nb}^* have in general obviously the same length and the theoretical arguments in Hall (1988) amount to a strong case against I_{nb}^* if $T(F)$ is a smooth functional of the mean of F; but it turns out that in case of $T(F) = F^{-1}(q)$ the ci I_{nb}^* is actually distribution free and has higher coverage probability than I_n^* (see Theorem 2.5 below). The integral probability transformation theorem implies for $T(F) = F^{-1}(q)$ the representation

$$G_n^* \left(n^{1/2}(F^{-1}(q) - F_n^{-1}(q)) \right) = H_n \left(F_n(F^{-1}(q)) \right), \tag{2.3}$$

where H_n denotes the df of the sample q-quantile pertaining to n independent and uniformly on $(0,1)$ distributed rvs, i.e. $H_n(x) = P\{\overline{F}_n^{-1}(q) \leq x\}$, $x \in \mathbb{R}$ (see Falk and Kaufmann (1989) for details).

Equation (2.3) implies

$$I_{nb}^* = \{\vartheta \in \mathbb{R} : \alpha \leq H_n(F_n(\vartheta)) < 1 - \alpha\} \tag{2.4}$$

$$= \{\vartheta \in \mathbb{R} : F_n^{-1}(H_n^{-1}(\alpha)) \leq \vartheta < F_n^{-1}(H_n^{-1}(1 - \alpha))\} = [X_{r_1:n}, X_{r_2:n}),$$

where $r_1 := r_1(n) := \langle nH_n^{-1}(\alpha)\rangle$, $r_2 := r_2(n) := \langle nH_n^{-1}(1 - \alpha)\rangle$, and $X_{1:n} \leq \ldots \leq X_{n:n}$ are the order statistics pertaining to $X_1, \ldots, X_n$. By $\langle x\rangle$ we denote the smallest integer not less than $x \in \mathbb{R}$. Consequently, I_{nb}^* is actually a classical ci for $F^{-1}(q)$ which uses order statistics (section 2.6.3 in the book by Serfling (1980)); it is therefore distribution free if only $F(F^{-1}(q)) = q$ and has level error of order $O(n^{-1/2})$. Note that this observation provides an explanation for the low level error of the ci I_n^*. The preceding observations are summarized in the following result (Lemma 5 and Corollary 6 in Falk and Kaufmann (1989), section 2.6.4 in Serfling (1980)).

2.5 Theorem. *Under the assumptions of Theorem 2.1 we have*

(i) $I_{nb}^* = [X_{r_1:n}, X_{r_2:n}),$

(ii) $n^{1/2} \left(P\{F^{-1}(q) \in I_{nb}\} - P\{F^{-1}(q) \in I_n\} \right)$

$$\to_{n\to\infty} 2\varphi(\Phi^{-1}(\alpha))(\Phi^{-1}(\alpha))^2/(q(1-q))^{1/2},$$

(iii) $n^{1/2}$ length of $I_{nb}^* = n^{1/2}$ length of $I_n^* = n^{1/2}(X_{r_2:n} - X_{r_1:n})$

$$\to_{n\to\infty} 2\Phi^{-1}(1-\alpha)(F^{-1})'(q)(q(1-q))^{1/2} \text{ almost sure.}$$

It is an open problem whether the bootstrap approach with backward critical points tries to pick up those order statistics as critical values of cis for $F^{-1}(q)$ which are optimal in some sense.

An extensive simulation study by Dohmann (1990)[1] shows that the theoretical advantage of I_{nb}^* over I_n^*, formulated in part (ii) of the preceding result, shows already up for small sample sizes n. This simulation study benefitted very much from the following representation of $G_n^{*-1}(\alpha)$ as a spacing of two particular order statistics which is immediate from formulae (2.2) and (2.4):

2.6 Lemma. *For any df F and any $q \in (0,1), \alpha \in (0,1)$ we have*

(i) $G_n^{*-1}(\alpha) = n^{1/2}(F_n^{-1}(H_n^{-1}(\alpha)) - F_n^{-1}(q)) = n^{1/2}(X_{r_1:n} - X_{(nq):n})$

(ii) $G_n^{-1}(\alpha) = n^{1/2}(F^{-1}(H_n^{-1}(\alpha)) - F^{-1}(q)).$

The practical advantage of Lemma 2.6 which is partially contained in Falk and Kaufmann (1989) and Hall and Martin (1989), lies in the fact that the representation of $G_n^{*-1}(\alpha)$ as the spacing of two particular order statistics avoids its computation by Monte Carlo approximation of G_n^*, which is usually inherent in the bootstrap method. This was already observed by Efron (1982).

Lemma 2.6 reveals in particular the following close relationship between the *inverse bootstrap process* $(W_n(\alpha))_{\alpha\in(0,1)} := (G_n^{*-1}(\alpha) - G_n^{-1}(\alpha))_{\alpha\in(0,1)}$ and the *empirical quantile process* $\beta_n(p) := n^{1/2}(F_n^{-1}(p) - F^{-1}(p))$, $p \in (0,1)$:

$$W_n(\alpha) = \beta_n(H_n^{-1}(\alpha)) - \beta_n(q), \ \alpha \in (0,1).$$

From the expansion $H_n^{-1}(\alpha) = q + (q(1-q))^{1/2}\Phi^{-1}(\alpha)n^{-1/2} + o(n^{-1/2})$ we obtain that the asymptotic behavior of W_n is determined by the local behavior of β_n. As a consequence it has been shown in Falk (1990b) that $n^{1/4}(W_n(\alpha))_{\alpha\in(0,1)}$ converges in distribution to a Gaussian process. From this result and Theorem 2.1 we see that the accuracy of the critical point $G_n^{*-1}(\alpha)$ is of order $O_P(n^{-1/4})$ whereas the level error of the pertaining ci is only of order $O(n^{-1/2})$.

In a forthcoming paper by Falk and Janas (1990) it is proved that the smoothed bootstrap approach entails *twosided* cis for I_n^* for $F^{-1}(q)$ which have level error of order $o(n^{-1/2})$, whereas *onesided* cis still have level error of order $O(n^{-1/2})$. Twosided smoothed bootstrap cis for the quantile therefore outperform competitors based on standard bootstrap, percentile method, bias correction and accelerated bias correction which have level error of precise order $O(n^{-1/2})$ in both onesided and twosided cases (Hall and Martin (1989)). This result underlines again the importance of a smoothed bootstrap.

---

[1]For an interactive PC version of the FORTRAN–programs underlying this study contact Birgit Dohmann, Fachbereich Mathematik, Universität GH Siegen, Hölderlinstr. 3, D–5900 Siegen 21.

References

Beran, R.J. (1987). *Prepivoting to reduce level error of confidence sets.* Biometrika 74, 457–468.

Dohman, B. (1990). *Confidence intervals for small sample sizes: Bootstrap vs. standard methods.* Diplom thesis, University of Siegen (in German).

Efron, B. (1979). *Bootstrap methods: another look at the jackknife.* Ann. Statist. 7, 1–26.

Efron, B. (1982). *The Jackknife, the Bootstrap and other Resampling Plans.* SIAM, Philadelphia.

Falk, M. (1983). *Relative efficiency and deficiency of kernel type estimators of smooth distribution functions.* Statist. Neerlandica 37, 73–83.

Falk, M. (1984). *Relative deficiency of kernel type estimators of quantiles.* Ann. Statist. 12, 261–268.

Falk, M. (1985). *Asymptotic normality of the kernel quantile estimator.* Ann. Statist. 13, 428–433.

Falk, M. (1990a). *Weak convergence of the maximum error of the bootstrap quantile estimate.* Statist. Probab. Letters, 10, 301–305.

Falk, M. (1990b). *Functional limit theorems for inverse bootstrap processes of sample quantiles.* Tentatively accepted for publication in Statist. Probab. Letters.

Falk, M. and Janas, D. (1990). *Edgeworth expansions for studentized and prepivoted sample quantiles with applications to confidence intervals.* Preprint.

Falk, M. and Kaufmann (1989). *Coverage probabilities of bootstrap confidence intervals for quantiles.* Ann. Statist., to appear.

Falk, M. and Reiss, R.-D. (1989a). *Weak convergence of smoothed and nonsmoothed bootstrap quantile estimates.* Ann. Probab. 17, 362–371.

Falk, M. and Reiss, R.-D. (1989b). *Bootstrapping the distance between smooth bootstrap and sample quantile distribution.* Probab. Th. Rel. Fields 82, 177-186.

Hall, P. (1988). *Theoretical comparison of bootstrap confidence intervals.* Ann. Statist. 16, 927–953.

Hall, P. and Martin, M.A. (1989). *A note on the accuracy of bootstrap percentile method confidence intervals for a quantile.* Statist. Probab. Letters 8, 197–200.

Hall, P., DiCiccio, T.J. and Romano, J.P. (1989). *On smoothing and the bootstrap.* Ann. Statist. 17, 692–704.

Jones, M.C. (1990). *The performance of kernel density functions in kernel distribution function estimation.* Statist. Probab. Letters 9, 129–132.

Reiss, R.-D. (1981). *Nonparametric estimation of smooth distribution functions.* Scand. J. Statist. 8, 116–119.

Reiss, R.-D. (1989). Approximate Distributions of order statistics (with Applications to Nonparametric Statistics). Springer Series in Statistics, Springer, New York.

Serfling, R.M. (1989). Approximation Theorems of Mathematical Statistics. Wiley, New York.

Silverman, B.W. (1986). Density Estimation (for Statistics and Data Analysis). Chapman and Hall, London.

Singh, K. (1981). *On the asymptotic accuracy of Efron's bootstrap.* Ann. Statist. 9, 1187–1195.

Bootstrapping Conditional Curves

M. Falk and R.-D. Reiss

Katholische Universität Eichstätt and Universität Gh Siegen

Abstract

It is shown that consistency and more refined results for the bootstrap
hold in the conditional framework if the analogue is valid in the unconditional case. We are particularly interested in the asymptotic performance
of bootstrap d.f.'s in case of mean and median regression functionals.

1 Introduction

Let T_0 be a functional from a space of univariate d.f.'s into the real line $I\!R$.
First, let us recall some basic facts from the unconditional set-up. A natural
estimator of the functional parameter $T_0(G)$ is the statistical functional $T_0(G_n)$
where G_n is the sample distribution function (d.f.) of n i.i.d. random variables
$W_1, \ldots, W_n$ with common d.f. G. The performance of that estimator may be
measured by the centered d.f.

$$T_{1,n}(G) \quad = \quad P_G\{T_0(G_n) - T_0(G) \leq \cdot\}, \tag{1}$$

thus, obtaining another functional having its values in the space of d.f.'s. Now,
Efron's [3] bootstrap approach may be adopted to obtain an estimator of the
d.f. $T_{1,n}(G)$. Consider the bootstrap d.f.

$$T_{1,n}(G_n) \quad = \quad P_{G_n}\{T_0(G_n^*) - T_0(G_n) \leq \cdot\} \tag{2}$$

where G_n^* is the sample d.f. based on the bootstrap sample $W_1^*, \ldots, W_n^*$ generated according to G_n.

For many functionals T_0 one can prove the weak consistency of the bootstrap
d.f.; that is, for every $\varepsilon > 0$,

$$P_G\{d(T_{1,n}(G_n), T_{1,n}(G)) \geq \varepsilon\} \quad \longrightarrow_{n \to \infty} \quad 0 \tag{3}$$

where $d(F, G) = \sup_{t \in I\!R} |F(t) - G(t)|$ denotes the Kolmogorov-Smirnov distance.

Moreover, much stronger results are valid in many cases; e.g. a non-degener-
ate limiting process of the bootstrap error is computed in [5] for the q-quantile
functional and the iterated bootstrap is studied in [6] (for a short summary see
[11]).

In analogy to the considerations above one may study conditional statistical functionals. Let $(X_1, Y_1), (X_2, Y_2), (X_3, Y_3), \ldots$ be i.i.d. random vectors that are replicas of the random vector (X, Y) having the bivariate d.f. F and density f. The density of X is denoted by g. Throughout this paper let $x \in I\!\!R$ be a fixed value such that $g(x) > 0$. Denote by

$$F(\cdot|x) \;=\; P(Y \leq \cdot | X = x) \tag{4}$$

the conditional d.f. of Y given $X = x$. The function $x \longrightarrow T_0(F(\cdot|x))$ is the mean and, respectively, the median regression function if T_0 is the mean and median functional. Generally, one may speak of a conditional curve.

The conditional sample d.f. $F_n(\cdot|x)$ may be defined by

$$F_n(t|x) \;=\; \frac{1}{K(n)} \sum_{i=1}^{n} 1_{(-\infty, t]}(Y_i) 1_{[x-a(n)/2, x+a(n)/2]}(X_i) \tag{5}$$

if

$$0 \;<\; K(n) \;=\; \sum_{i=1}^{n} 1_{[x-a(n)/2, x+a(n)/2]}(X_i), \tag{6}$$

where $a(n)$ is a bandwidth satisfying $a(n) \downarrow 0$ as $n \to \infty$. Notice that $F_n(\cdot|x)$ is a sample d.f. based on the Y_i belonging to those X_i lying in the interval $[x - a(n)/2, x + a(n)/2]$.

In analogy to (1), the performance of the conditional statistical functional $T_0(F_n(\cdot|x))$ as an estimator of $T_0(F(\cdot|x))$ will be measured by the d.f.

$$T_{2,n}(F) \;=\; P_F\{T_0(F_n(\cdot|x)) - T_0(F(\cdot|x)) \leq \cdot\}. \tag{7}$$

Notice that in (7) the probability is taken under a bivariate d.f. F whereas in (1) the underlying d.f. is univariate. Intuitively speaking, our target d.f. $T_{2,n}(F)$ is closely related to $T_{1,K(n)}(F(\cdot|x))$. Since $F_n(\cdot|x)$ is an estimator of $F(\cdot|x)$ one may resample according to the conditional sample d.f. $F_n(\cdot|x)$ to obtain an estimator of the d.f. $T_{2,n}(F)$. The conditional bootstrap d.f. is given by

$$\begin{aligned}
T_{1,K(n)}&(F_n(\cdot|x)) \\
&= \; P_{F_n(\cdot|x)}\{T_0(G^{**}_{K(n)}) - T_0(F_n(\cdot|x)) \leq \cdot\}
\end{aligned} \tag{8}$$

where $G^{**}_{K(n)}$ is the sample d.f. based on the bootstrap sample

$$W_1^{**}, \ldots, W_{K(n)}^{**}$$

generated according to the d.f. $F_n(\cdot|x)$. Notice that, in fact, $T_{1,K(n)}(F_n(\cdot|x))$ is a d.f. of the form (1) with n replaced by the random sample size $K(n)$ and G replaced by $F_n(\cdot|x)$.

We are interested in the performance of $T_{1,K(n)}(F_n(\cdot|x))$ as an estimator of $T_{2,n}(F)$. It is the aim of the present paper to prove the weak consistency

$$P_F\{d(T_{1,K(n)}(F_n(\cdot|x)), T_{2,n}(F)) \geq \varepsilon\} \longrightarrow_{n\to\infty} 0, \ \varepsilon > 0, \tag{9}$$

and stronger results if corresponding results hold in the unconditional set-up.

2   The Main Results

Our basic condition will be that the statistical functional $T_0(G_n)$ is asymptotically normal, where G_n is the sample d.f. based on k i.i.d. random variables with common d.f. $G = F(\cdot|x)$. More precisely, suppose that for some $\sigma > 0$, $\delta \in [0, 1/2]$, and $C > 0$,

$$d\left(T_{1,n}(F(\cdot|x)), \Phi(\cdot\, n^{1/2}/\sigma)\right) \leq C n^{-\delta}, \ n \in I\!N, \tag{10}$$

where Φ denotes the standard normal d.f. Notice that this condition is satisfied for the mean and median functional with $\delta = 1/2$ under mild conditions imposed on $F(\cdot|x)$.

Theorem 1 *Suppose that, in addition to the conditions (10) and (19) (cf. Theorem 4), weak consistency*

$$P_{F(\cdot|x)}\{d(T_{1,k}(G_k), T_{1,k}(F(\cdot|x))) \geq \varepsilon\} \longrightarrow_{k\to\infty} 0, \ \varepsilon > 0, \tag{11}$$

of the unconditional bootstrap d.f. holds.
   Then, if $na(n) \to \infty$ and $na(n)^{3+2\beta} \to 0$ as $n \to \infty$, weak consistency (9) of the conditional bootstrap d.f. holds.

REMARK. From (10) it is immediate that (11) is equivalent to

$$P_{F(\cdot|x)}\{d\left(T_{1,k}(G_k), \Phi(\cdot\, k^{1/2}/\sigma)\right) \geq \varepsilon\} \longrightarrow_{k\to\infty} 0, \ \varepsilon > 0. \tag{12}$$

In that context we also refer to [8] where such a question is treated in the particular case of the mean value functional.

A consistency result for the bootstrap d.f. in case of the regression mean was obtained by Dikta [2] in the context of nearest neighborhood regression estimates.
   By strengthening condition (12) we will obtain a stronger result for the bootstrap d.f. in the conditional framework.

Theorem 2 *Suppose, in addition to conditions (10) and (18), that for every $u \in I\!R$,*

$$P_{F(\cdot|x)}\{A(k)\, d(T_{1,k}(G_k), \Phi(\cdot\, k^{1/2}/\sigma)) \leq u\} \longrightarrow_{k\to\infty} H(u) \tag{13}$$

where H is a continuous d.f. and A is a strictly positive function that is regularly varying at infinity.

Then, if $na(n) \to \infty$ and $na(n)^{3+2\beta} \to 0$ as $n \to \infty$,

$$P_F\{A(na(n)g(x))\, d(T_{1,K(n)}(F_n(\cdot|x)),\ \Phi(\cdot\ K(n)^{1/2}/\sigma)) \le u\} \to_{n\to\infty} H(u)$$

for every $u \in \mathbb{R}$.

Next we verify condition (14) for regression quantiles and the regression mean.

EXAMPLE 1. Let $T_0(G) = G^{-1}(q)$ be the q-quantile of the d.f. G. Under milder conditions it is proved by Falk [4] that (13) holds for $A(k) = k^{1/4}$, $\sigma = (q(1-q))^{1/2}/v(z_0)$, and

$$H(u) = \left(P\left\{ \sup_{t\ge 0} |\psi(t)\, B(t)| \le u \right\} \right)^2, \ u \ge 0,$$

where $v(z) = \partial F(z|x)/\partial z$, $z_0 = F(\cdot|x)^{-1}(q)$,

$$\psi(t) = (v(z_0)/(q(1-q)))^{1/2}\Phi'\left(tv(z_0)/(q(1-q))^{1/2}\right)$$

and B is the standard Brownian motion on $\mathbb{R}_+$.

EXAMPLE 2. Let $T_0(G) = \int z\, dG(z)$ be the mean of G. Utilizing Theorem 1, (1.4), in Singh [12] one may prove that (13) holds for $A(k) = k^{1/2}$, $\sigma^2 = \int(z-\mu)^2\, F(dz|x)$ and $H(u) = P\{W \le u\}$ where $\mu = \int z\, F(dz|x)$ and W is a r.v. given by

$$W = \sup_{t\in \mathbb{R}} \Phi'(t)|t(\sigma_4/(2\sigma^2))Z + \mu_3(1-t^2)/(6\sigma^3)|$$

with Z being a standard normal r.v., $\mu_3 = \int(z-\mu)^3\, F(dz|x)$, and $\sigma_4^2 = \int((z-\mu)^2 - \sigma^2)^2\, F(dz|x)$.

Our present result may easily be extended to the question of the simultaneous treatment of regression bootstrap d.f.'s given x_i for several points x_i (see also [7]).

Corresponding results can be established in case of smooth bootstrap. In that case, the sample d.f. G_n has to be replaced by the smooth sample d.f. of the form

$$G_{n,\alpha(n)}(t) \ = \ \int U\left(\frac{t-z}{\alpha(n)}\right) G_n(dz) \tag{14}$$

where U is a d.f. and $\alpha(n) \downarrow 0$ as $n \to \infty$. Notice that $G_{n,\alpha(n)}$ and, hence, $T_{1,n}(G_{n,\alpha(n)})$ is a functional of the sample d.f. G_n.

Moreover, the conditional d.f. $F_n(\cdot|x)$ has to be replaced by its smooth version

$$F_{n,\alpha(K(n))}(t|x) \;=\; \int U\left(\frac{t-z}{\alpha(K(n))}\right) F_n(dz|x), \; t \in I\!R. \tag{15}$$

3 Auxiliary Results and Proofs

The conditional d.f. $F_n(\cdot|x)$ in (5) is closely related to the truncated empirical point process

$$N_n(\cdot|x) \;\stackrel{\cdot}{=}\; \sum_{i=1}^{n} \varepsilon_{Y_i}(\,\cdot\,)\varepsilon_{X_i}([x - a(n)/2, x + a(n)/2]) \tag{16}$$

where $\varepsilon_z(B) = 1_B(z)$. In particular, a functional of $F_n(\cdot|x)$ may be regarded as a functional of $N_n(\cdot|x)$.

Consider the Poisson point process

$$N_n^*(\cdot|x) \;=\; \sum_{i=1}^{\tau(n)} \varepsilon_{W_i} \tag{17}$$

where $\tau(n), W_1, W_2, \ldots$ are independent r.v.'s such that $W_1, W_2, \ldots$ are identically distributed with common d.f. $F(\cdot|x)$ as defined in (4), and $\tau(n)$ is a Poisson r.v. with expectation $na(n)g(x)$ where g is the density of X.

Denote by H the Hellinger distance. Recall that the Hellinger distance dominates the variational distance. From [7] we quote the following two theorems.

If a Berry-Esséen type theorem is valid for the unconditional statistical functional (that is condition (10)) then a similar result also holds for the conditional one.

Theorem 3 *If (10) holds then, with $T_{2,n}(F)$ as in (7),*

$$d\left(T_{2,n}(F), \; \Phi(\cdot (na(n)g(x))^{1/2}/\sigma)\right)$$
$$\leq \quad D\left(na(n)\right)^{-\delta} + H(\mathcal{L}(N_n(\cdot|x)), \mathcal{L}(N_n^*(\cdot|x)))$$

where $D > 0$ only depends on C.

A bound for the Hellinger distance $H(\mathcal{L}(N_n(\cdot|x)), \mathcal{L}(N_n^*(\cdot|x)))$ is obtained in the following theorem.

Theorem 4 *If for ρ in a neighborhood of 0*

$$f(x+\rho, y)^{1/2} \;=\; f(x,y)^{1/2}\left(1 + \rho h(y) + O(|\rho|^{1+\beta} r(y))\right) \tag{18}$$

for some $0 < \beta \leq 1$, and functions h and r satisfying

$$\int (h(y)^4 + r(y)^4) \ f(x,y) \, dy < \infty,$$

then

$$H(\mathcal{L}(N_n(\cdot|x)), \mathcal{L}(N_n^*(\cdot|x))) = O(a(n) + n^{1/2}a(n)^{(3+2\beta)/2}).$$

In the present paper the replacement of $N_n(\cdot|x)$ by $N_n^*(\cdot|x)$ is essential. Alternatively, one may adopt the method of poissonization where the empirical process $\sum_{i=1}^n \varepsilon_{(X_i,Y_i)}$ is replaced by the Poisson process $\sum_{i=1}^{\tau(n)} \varepsilon_{(X_i,Y_i)}$ with $\tau(n)$ denoting a Poisson r.v. with expectation n. Notice that

$$P\{s^{-1/2}|\tau(s) - s| \geq \varepsilon\} \leq 2e^{-\varepsilon^2/4} \tag{19}$$

for $s > 0$ and $\varepsilon > 0$ (cf. [1]). Poissonization was successfully applied to prove consistency results for estimators of conditional curves (cf. [9], [10], [13]). One may conjecture that corresponding bootstrap results for the bootstrap may as well be proved by adopting that method. Due to the large deviation between $\tau(n)$ and n it is doubtful that stronger results (like that in Theorem 2) can be established by using poissonization.

Theorems 3 and 4 will be the basic tools for proving Theorems 1 and 2.

PROOF OF THEOREM 1. Put $b(n) = na(n)g(x)$. From Theorems 3 and 4 we deduce for sufficiently large n,

$$
\begin{aligned}
P_F&\{d(T_{1,K(n)}(F_n(\cdot|x)), T_{2,n}(F)) \geq \varepsilon\} \tag{20}\\
&\leq \ P_F\{d(T_{1,K(n)}(F_n(\cdot|x)), \Phi(\cdot \, b(n)^{1/2}/\sigma)) \\
&\qquad + d(\Phi(\cdot \, b(n)^{1/2}/\sigma), T_{2,n}(F)) \geq \varepsilon\} \\
&\leq \ P_F\{d(T_{1,K(n)}(F_n(\cdot|x)), \Phi(\cdot \, b(n)^{1/2}/\sigma)) \geq \varepsilon/2\} \\
&= \ \int P_{F(\cdot|x)}\{d(T_{1,k}(G_k), \Phi(\cdot \, b(n)^{1/2}/\sigma)) \geq \varepsilon/2\} \, \mathcal{L}(\tau(n))(dk) + o(n^0)
\end{aligned}
$$

where the last step is achieved by replacing the truncated empirical process $N_n(\cdot|x)$ by $N_n^*(\cdot|x)$ (cf. Theorem 3) and by conditioning on the Poisson r.v. $\tau(n)$ with expectation $b(n)$. Moreover,

$$
\begin{aligned}
P_{F(\cdot|x)}&\{d(T_{1,k}(G_k), \Phi(\cdot \, b(n)^{1/2}/\sigma)) \geq \varepsilon/2\} \tag{21}\\
&\leq \ P_{F(\cdot|x)}\{d(T_{1,k}(G_k), \Phi(\cdot \, k^{1/2}/\sigma)) \\
&\qquad + d(\Phi(\cdot \, k^{1/2}/\sigma), \Phi(\cdot \, b(n)^{1/2}/\sigma)) \geq \varepsilon/2\} \\
&\leq \ f(k) + 1_{\{m:\, d(\Phi(\cdot \, m^{1/2}/\sigma),\ \Phi(\cdot \, b(n)^{1/2}/\sigma)) \geq \varepsilon/4\}}(k)
\end{aligned}
$$

where

$$f(k) = P_{F(\cdot|x)}\{d(T_{1,k}(G_k), \Phi(\cdot \, k^{1/2}/\sigma)) \geq \varepsilon/4\} \to_{k \to \infty} 0$$

according to condition (12).

Combining (20) and (21), and applying the Markov inequality we obtain

$$P_F\{d(T_{1,K(n)}(F_n(\cdot|x)),\ T_{2,n}(F)) \geq \varepsilon\} \tag{22}$$

$$\leq \int f(k)\mathcal{L}(\tau(n))(dk)$$

$$+ \frac{4}{\varepsilon}\int d(\Phi(\cdot\ k^{1/2}/\sigma),\ \Phi(\cdot\ b(n)^{1/2}/\sigma))\ \mathcal{L}(\tau(n))(dk) + o(n^0).$$

It is immediate from (19) that the first integral converges to zero. To prove the convergence to zero of the second integral one may use arguments of the proof of Lemma 1 in [7].

PROOF OF THEOREM 2. In analogy to the proof of Theorem 1 we obtain for every $u \in I\!R$,

$$P_F\{A(na(n)g(x))\ d(T_{1,K(n)}(F_n(\cdot|x)),\ \Phi(\cdot\ K(n)^{1/2}/\sigma)) \leq u\} - H(u)$$

$$= \int\{P_{F(\cdot|x)}\{A(na(n)g(x))\ d(T_{1,k}(G_k),\ \Phi(\cdot\ k^{1/2}/\sigma)) \leq u\}$$

$$- H(u)\}\ \mathcal{L}(\tau(n))(dk) + o(n^0) \tag{23}$$

$$= \int\left(H\left(\frac{A(k)}{A(na(n)g(x))}\ u\right) - H(u)\right)\mathcal{L}(\tau(n))(dk) + o(n^0)$$

where the last step follows from condition (13). W.l.g. we may restrict the integral to the interval $[c(n),\ d(n)]$ where $c(n) = na(n)g(x) - (na(n))^{3/4}$ and $d(n) = na(n)g(x) + (na(n))^{3/4}$. (cf. (19)). Moreover,

$$\sup_{c(n)\leq k\leq d(n)} |k/(na(n)g(x)) - 1\ | \to 0,\ n \to \infty,$$

and, hence, the uniform convergence theorem for regularly varying functions and the continuity of H yield

$$\sup_{c(n)\leq k\leq d(n)} |H\left(\frac{A(k)}{A(na(n)g(x)}\ u\right) - H(u)\ | \to 0, n \to \infty. \tag{24}$$

Combining (23) and (24) we get the desired conclusion.

References

[1] Devroye, L.P. and Györfi, L. (1985). *Nonparametric density estimation.* New York: Wiley.

[2] Dikta, G. (1990). Bootstrap approximation of nearest neighborhood regression function estimates. J. Mult. Analysis **32** 213–229.

[3] Efron, B. (1979). Bootstrap methods: Another look at the jackknife. Ann. Statist. **7** 1–26.

[4] Falk, M. (1990). Weak convergence of the maximum error of the bootstrap quantile estimate. Statist. Probab. Letters **10** 301–305.

[5] Falk, M. and Reiss, R.-D. (1989). Weak convergence of smoothed and non-smoothed bootstrap quantile estimates. Ann. Probab. **17** 362–371.

[6] Falk, M. and Reiss, R.-D. (1989). Bootstrapping the distance between smooth bootstrap and sample quantile distribution. Probab. Th. Rel. Fields **82** 177–186.

[7] Falk, M. and Reiss, R.-D. (1989). Statistical Inference of Conditional Curves: Poisson Process Approach. Preprint 231, University of Siegen.

[8] Giné, E. and Zinn, J. (1989). Necessary conditions for the bootstrap of the mean. Ann. Statist. **17** 684–691.

[9] Härdle, W. and Kelly, G. (1987). Nonparametric kernel regression estimation—optimal choice of bandwidth. Statistics **18** 21–35.

[10] Härdle, W., Janssen, P., and Serfling, R. (1988). Strong uniform consistency rates for estimators of conditional functionals. Ann. Statist. **16** 1428–1449.

[11] Reiss, R.-D. (1989). *Approximate Distributions of Order Statistics: With Applications to Nonparametric Statistics.* Springer Series in Statistics. New York: Springer.

[12] Singh, K. (1981). On the asymptotic accuracy of Efron's bootstrap. Ann. Statist. **9** 1187–1195.

[13] Zhao, L.C. (1989). Exponential bound of mean error for the kernel estimates of regression functions. J. Mult. Analysis **29** 260–273.

Resampling Stochastic Processes using a Bootstrap Approach

Anders Nordgaard

Department of Mathematics, Linköping Institute of Technology

S-581 83 Linköping, SWEDEN

Abstract

A bootstrap method for weakly stationary Gaussian sequences is presented. The resampling is done among the coefficients of the random Fourier representation of a sample from the sequence. The independent increments for the spectral representation of a sequence justifies the choice of the method and the resampling provides new bootstrap realizations. Validity of the method is checked on a typical spectral estimator and a correlation estimator using computer simulation of large samples. The results are satisfactory and point towards asymptotic validity of the method.

1. Resampling procedure This article is a short version of the thesis "On the Resampling of Stochastic Processes using a Bootstrap Approach" (Nordgaard (1990)). We present a resampling method related to Efron's Bootstrap (Efron (1979)), which is analyzed using computer simulation.

Let $\{X_t\}_{t=-\infty}^{\infty}$ be a stationary, real-valued, discrete-time Gaussian stochastic process with mean $EX_t \equiv 0$ and covariance function $R(\tau) = EX_{t+\tau}X_t$. Consider a sample $\{x_t\}_{t=-n}^{n}$ from the process and call this sample a *signal*. Assume that the *spectral representation*

$$X_t = \int_{-\pi}^{\pi} e^{i\omega t}\, dZ(\omega) \tag{1}$$

exists, where $Z(\omega)$, $-\pi \leq \omega \leq \pi$ is a stochastic process with zero mean, orthogonal increments and $E\,|Z(\omega_2) - Z(\omega_1)|^2 = F(\omega_2) - F(\omega_1)$, $\omega_1 \leq \omega_2$, where F is the spectral distribution function. The sample spectral representation

$$x_t = \sum_{k=-n}^{n} a_{k,n} e^{i\omega_k t}, \tag{2}$$

where $a_{k,n} = \frac{1}{2n+1}\sum_{t=-n}^{n} x_t e^{-i\omega_k t}$ and $\omega_k = k2\pi/(2n+1)$, may be thought of as an approximation of (1) in writing $x_t = \sum_{k=-n}^{n} a_{k,n} e^{i\omega_k t} = \int_{-\pi}^{\pi} e^{i\omega t}\, dZ_n(\omega)$, where $Z_n(\omega) = \sum_{k:\omega_k \leq \omega} a_{k,n}$, $-\pi \leq \omega \leq \pi$. For brevity, we will write a_k instead of $a_{k,n}$. It can be shown that $E\,|dZ_n(\omega) - dZ(\omega)|^2 \to 0$ as $n \to \infty$ (Rosenblatt(1985), p. 21). If $\sum_{\tau=-\infty}^{\infty} R(\tau) < \infty$ the *spectral density function* $S(\omega) = \frac{dF(\omega)}{d\omega}$ exists and it may be verified that (Rosenblatt(1985), p 126–127)

$$\frac{2n+1}{2\pi} E\,|a_k|^2 = \frac{1}{2\pi}\sum_{\tau=-2n}^{2n} R(\tau)\left(1 - \frac{|\tau|}{2n+1}\right)e^{-i\omega_k \tau} \to S(\omega_k) \quad \text{as} \quad n \to \infty. \tag{3}$$

The indicated similarity between a_k and $\Delta Z_k = Z(\omega_k + \frac{1}{2}\Delta\omega) - Z(\omega_k - \frac{1}{2}\Delta\omega)$ gives hope for weak dependency between different a_k. Because of the Gaussian distribution $Z(\omega)$ has independent increments. If $S(\omega)$ is smooth the coefficients $\{a_{k-1}, a_k, a_{k+1}\}$ may be treated as approximately i.i.d. for moderately long signals and the bootstrap might be applicable to such sets. However, we extend the procedure by separating the *amplitude* $|a_k|$ and the *phase* $\arg a_k$ and do the resampling independently for each of these two. The whole set of coefficients $\{a_k\}_{k=-n}^{n}$ may be denoted by a. A resampled coefficient is written as a_k^* and the corresponding amplitude and phase as $|a|_k^*$ and $(\arg a)_k^*$ respectively. We will use the notation

$$E_*(\cdot) = E(\cdot\,|a) \quad \text{and} \quad P_*(\cdot) = P(\cdot\,|a),$$

when we make inference about the resampled quantities. When X_t is real-valued, $a_{-k} = \overline{a_k}$, and only a_k $(k \geq 0)$ needs to be resampled. The elements of the set $\{|a_{k-1}|, |a_k|, |a_{k+1}|\}$ are asymptotically independent (Rosenblatt (1985), p. 131) and a_{k-1}, a_k, a_{k+1} are Gaussian with mean zero and almost equal variances when $S(\omega)$ is smooth and n is not too small. The amplitude $|a|_k^*$ may then be drawn with equal weights from the set $\{|a_{k-1}|, |a_k|, |a_{k+1}|\}$ when $0 < k < n$. For the cases $k = 0$ and $k = n$ one might reduce the resampling set to two coefficients but this will give us some extra terms when we expand the signal size towards infinity (Nordgaard (1990), p. 18–20). Instead we let $|a|_0^* = |a_0|$ with probability 1/3 and $|a|_0^* = |a_1|$ with probability 2/3. This may be thought of as adding the term $|a_{-1}|$ to the resampling set $\{|a_0|, |a_1|\}$ as $|a_{-1}| = |a_1|$. For $|a|_n^*$ we give the weights 1/3 to $|a_{n-1}|$ and 2/3 to $|a_n|$. The cyclic behaviour of the discrete Fourier transform implies that the "added term" to this resampling set should be $|a_{-n}|$ which is equal to $|a_n|$. We may write $a_k = b_k + ic_k$ where $b_k = \sum_{t=-n}^{n} X_t \cos \omega_k t$ and $c_k = \sum_{t=-n}^{n} X_t \sin \omega_k t$. Then it is easily shown that $E\, b_k = E\, c_k = 0$ and $E\, b_k^2 \simeq E\, c_k^2$ if $k \neq 0$ (asymptotically $E\, b_k^2 = E\, c_k^2$ (Rosenblatt (1985), p. 130–131)). Further it may also be verified that $E\, b_k c_l = 0$ (Hjorth (1987), p. 92). Since X_t is a Gaussian sequence, b_k and c_k are independent and approximately equally distributed Gaussian variables. Thus a_k is (approximately) rotation-invariant. A semiparametric approach may then be to draw $(\arg a)_k^*$ uniformly on the interval $(-\pi, \pi)$. However, if we instead of the above approach resample the quantities b_k and c_k on basis of the earlier discussion we will combine the information in $|a_k^*|$ and $(\arg a)_k^*$ in a suitable way. We suggest the following:

Draw ϕ_1 and ϕ_2 independently from a $Re(-\pi, \pi)$-distribution and let $e^{i(\arg a)_k^*} = \cos \phi_1 + i \sin \phi_2$. This procedure implies that $E_* e^{i(\arg a)_k^*} = 0 = E_*(e^{i(\arg a)_k^*})^2$. As a result the procedure will in general change the amplitude , but the mean effect contribution is still preserved as: $E_* |a_k^*|^2 = E_* \||a|_k^*(\cos \phi_1 + i \sin \phi_2)|^2 = E_* \||a|_k^*|^2 E_*(\cos^2 \phi_1 + \sin^2 \phi_2) = E_*(|a|_k^*)^2$. It is clear that $E_* a_k^* = E_* |a|_k^* E_* e^{i(\arg a)_k^*} = 0 \Rightarrow E_* X_t^* \equiv 0$ and that $E_*(a_k^*)^2 = 0$. Further (See Nordgaard (1990), p. 11–12,15–18) we will have: $R_*(t + \tau, t) = E_* X_{t+\tau}^* \overline{X_t^*} = \sum_{k=-n}^{n} \sum_{l=-n}^{n} E_* a_k^* \overline{a_l^*} e^{i\omega_k(t+\tau)} e^{-i\omega_l t} = \sum_{k=-n}^{n} E_* |a_k^*|^2 e^{i\omega_k \tau}$, as a_k^*, a_l^* are independent quantities when $|k| \neq |l|$ and $E_*(a_k^*)^2 = 0$. It is also easily verified that $\sum_{k=-n}^{n} E |a_k|^2 e^{i\omega_k \tau} = R(\tau)(1 - \frac{|\tau|}{2n+1}) \rightarrow R(\tau)$ as $n \rightarrow \infty$, for the signal $\{X_t\}_{t=-n}^{n}$. By developing $E_* |a_k^*|^2$ according to the amplitude resampling method suggested above, we will get

$$R_*(\tau) \simeq \sum_{k=-n}^{n} |a_k|^2 e^{i\omega_k \tau} = T_n(\tau) \tag{4}$$

for large n. (A few extra terms will vanish. See Nordgaard (1990, p. 20–21).) It is seen that $T_n(\tau)$ is an unbiased estimator of $\sum_{k=-n}^{n} E|a_k|^2 e^{i\omega_k \tau}$ and thus an asymptotically unbiased estimator of $R(\tau)$. From the discussions preceding (3) and a standard result for Gaussian variables it may be shown that (Nordgaard (1990), p. 21–22) $Var T_n(\tau) \rightarrow 0$ as $n \rightarrow \infty$ and thus $E(T_n(\tau) - R(\tau))^2 \rightarrow 0$.

To summarize we have found a resampling procedure that generates resamples or "resignals" with covariance structure converging in $\mathcal{L}^2$ to the covariance function of the basic sequence. The advantages of the method are (i) it does not depend on any linear model-description for the stochastic process and (ii) the resampled realizations may be used in estimation problems concerning the whole sample of a stochastic process. In the next section we will show that the method performs well on a couple of typical estimation problems by using computer simulation of long signals.

2. Simulation studies To analyze the actual advantages and/or disadvantages of the resampling procedures derived in section one it is somewhat necessary to use effective computer simulations. These can easily be applied to well-defined estimators, $\hat{\theta}$ in both frequency and time domain. We have developed a FORTRAN program for the supercomputer CRAY X-MP/48 because of the need for heavy calculations. This is so because the purpose of these simulations was to show on the average a good performance of the method for long signals. Applications on more "normal" signal lengths are easily implemented on most computers. To make inference on the statistic $T = \hat{\theta} - \theta$, we want to apply the bootstrap principle and compute the statistic $T^* = \hat{\theta}^* - \tilde{\theta}$ on the resampled signal, where $\tilde{\theta}$ is the corresponding parameter in the empirical distribution induced by the resampling. Consider a so-called *spectrograph estimator* $\hat{\theta} = \sum_{k=-n}^{n} |a_k|^2 W(\omega - \omega_k)$, where $W(\omega)$ is weight function sampled on $2n + 1$ points. By (3) we have

$$\lim_{n \to \infty} E\hat{\theta} = \int_{-\pi}^{\pi} W(\omega - \lambda) \, dF(\lambda). \tag{5}$$

We will use (5) as our defined parameter θ. Remembering that we could write $x_t = \int_{-\pi}^{\pi} e^{i\omega t} \, dZ_n(\omega)$ with $Z_n(\omega) = \sum_{k:\omega_k \leq \omega} a_k$, a corresponding spectral distribution function $F_n(\omega)$, which satisfies: $F_n(\omega_2) - F_n(\omega_1) = E|Z_n(\omega_2) - Z_n(\omega_1)|^2 = \sum_{k:\omega_1 < \omega_k \leq \omega_2} E|a_k|^2$ for $\omega_1 < \omega_2$, can be defined and thus $dF_n(\omega) = \begin{cases} E|a_k|^2 & \omega = \omega_k \\ 0 & \omega \neq \omega_k \end{cases}$. Replacing $dF(\lambda)$ with $dF_n(\lambda)$ in (5) we will have $\int_{-\pi}^{\pi} W(\omega - \lambda) \, dF_n(\lambda) = \sum_{k=-n}^{n} W(\omega - \omega_k) E|a_k|^2$, and for the resampled signal we get $\tilde{\theta} = \sum_{k=-n}^{n} W(\omega - \omega_k) E_* |a_k^*|^2$. We also consider the estimator of correlation as $\hat{\rho}(\tau) = \hat{R}(\tau)/\hat{R}(0)$, where $\hat{R}(\tau) = \sum_{t=-n}^{n-\tau} x_{t+\tau} x_t$. Clearly $\tilde{R}(\tau) = R_*(\tau)$ and $\tilde{\rho}(\tau) = \tilde{R}(\tau)/\tilde{R}(0)$.

The resampling procedure $\mathcal{R}$, is applied 200 times on each of 400 independently generated signals. The 400 basic signals will represent the situations appearing "on the average". On the 200 resampled signals from one particular basic signal the statistic T^* is computed for each of the estimators described above. For one estimator this will give 200 resampling estimates and an empirical distribution function $\hat{\mathcal{F}}^*$ may be calculated. As a first measure of goodness the 5-, 20-, 50-, 80- and 95-percentiles of $\hat{\mathcal{F}}^*$ are computed. For all 400 basic signals we then get empirical distributions for each of these percentiles. We put these distributions in relation to the empirical distribution of T (and its percentiles) computed on the basic signals. The graphical illustration of this comparison is shown for each estimator which gives an indication of the performance of the resampling. A second measure illustrates how well significance levels are achieved when testing hypothesis about the parameter θ. The quantity $\hat{\mathcal{F}}^*_{\hat{\theta}^* - \tilde{\theta}}(\hat{\theta} - \theta)$, where $\hat{\mathcal{F}}^*_{\hat{\theta}^* - \tilde{\theta}}$ is the empirical distribution function based on the 200 resampled values $\hat{\theta}^*$ for each basic signal, should be distributed uniformly on $(0, 1)$ if the resampling is successful. This measure is more comprehensively described in Nordgaard (1990, p. 53–54). The quantity is computed for each of the 400 resampling distributions and the values are shown in a histogram for each estimator. The value of the Chi-square Statistic for a test of uniformly distributed values is also computed.

Signals with 501 points are generated from the model

$$X_t = 0.7 X_{t-1} - 0.2 X_{t-2} + \varepsilon_t, \quad \varepsilon_t \text{ i.i.d.} N(0, 2).$$

We use a Gaussan probability density function with $\sigma = 0.5$ as weight function and estimate $S(\omega_0) = S(0)$. The correlation function is estimated for $\tau = 1$. The performances are shown for the spectral density estimator in figures 1 and 2, and for the correlation estimator in figures 3 and 4. The values are listed in table 1.

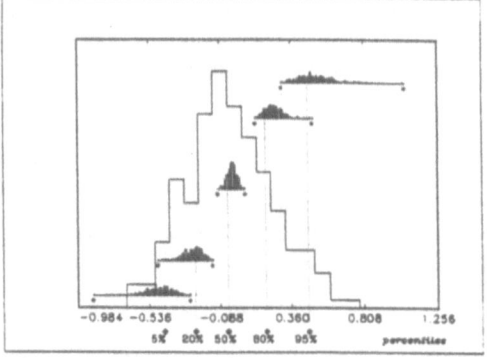

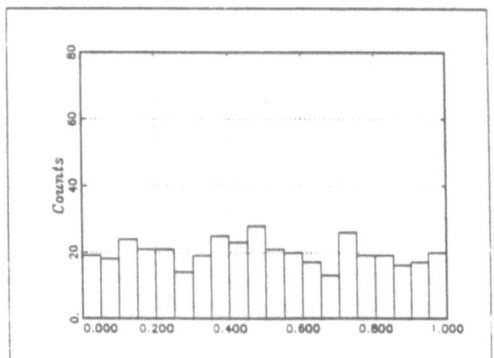

Figure 1: Histogram over 400 estimates of $S_b(\omega_0)$ (with Gaussian smoothing filter) , completed with histograms over percentiles computed on the resampling distributions based on 200 re-samplings on each basic signal.

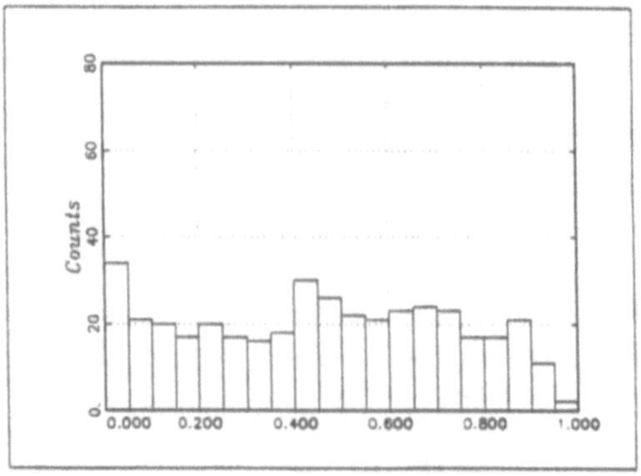

Figure 2: Histogram over 400 computed values of $\hat{\mathcal{F}}_{\hat{S}_b^*(\omega_0)-\tilde{S}_b(\omega_0)}(\hat{S}_b(\omega_0) - S_b(\omega_0))$ in the re-sampling shown in fig. 1. Value of Chi-Square Statistic is 41.7

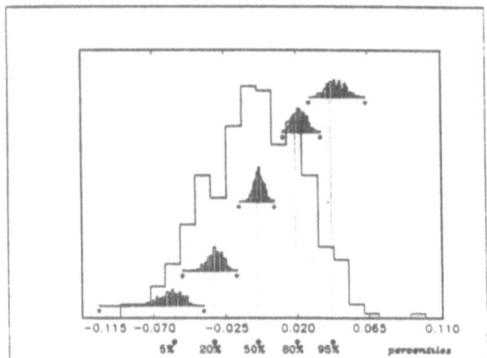

Figure 3: Histogram over 400 estimates of $\rho(1)$, completed with histograms over per-centiles computed on the resampling distributions based on 200 resamplings on each basic signal.

Figure 4: Histogram over 400 computed values of $\hat{\mathcal{F}}_{\hat{\rho}^*(1)-\tilde{\rho}(1)}(\hat{\rho}(1) - \rho(1))$ in the resampling shown in fig. 3. Value of Chi-Square Statistic is 14.0

185

Signal: $x_{-250}, \ldots, x_{250}$ from $X_t = 0.7X_{t-1} - 0.2X_{t-2} + \varepsilon_t$, $\varepsilon_t \sim \text{iid} N(0,2)$				
400 × 200 resamples				
statistic	average	median		stand.err.
$\tilde{S}_b(\omega_0) - S_b(\omega_0)$ $(\sigma = 0.5)$	-0.0244	-0.0436		0.2755
perc: 5: -0.4399, 20: -0.2481, 80: 0.1989 95: 0.4643				
5%perc$(\tilde{S}_b^*(\omega_0) - \tilde{S}_b(\omega_0))$	-0.4866	-0.4804		0.0939
20%perc$(\tilde{S}_b^*(\omega_0) - \tilde{S}_b(\omega_0))$	-0.2646	-0.2564		0.0587
50%perc$(\tilde{S}_b^*(\omega_0) - \tilde{S}_b(\omega_0))$	-0.0156	-0.0151		0.0284
80%perc$(\tilde{S}_b^*(\omega_0) - \tilde{S}_b(\omega_0))$	0.2537	0.2494		0.0581
95%perc$(\tilde{S}_b^*(\omega_0) - \tilde{S}_b(\omega_0))$	0.5168	0.4968		0.1164
$\hat{\rho}(1) - \rho(1)$	-0.0061	-0.0052		0.0299
perc: 5: -0.0570, 20: -0.0323, 80: 0.0193 95: 0.0413				
5%perc$(\hat{\rho}^*(1) - \hat{\rho}(1))$	-0.0598	-0.0589		0.0082
20%perc$(\hat{\rho}^*(1) - \hat{\rho}(1))$	-0.0315	-0.0312		0.0053
50%perc$(\hat{\rho}^*(1) - \hat{\rho}(1))$	-0.0041	-0.0040		0.0034
80%perc$(\hat{\rho}^*(1) - \hat{\rho}(1))$	0.0222	0.0220		0.0043
95%perc$(\hat{\rho}^*(1) - \hat{\rho}(1))$	0.0452	0.0446		0.0065

Table 1: Results from the resampling of 400 signals of size 501 points from the model $X_t = 0.7X_{t-1} - 0.2X_{t-2} + \varepsilon_t$, ε_t iid $N(0,2)$

The figures and the tables show that the resamples of the realizations give reasonable results for the studied estimators. The results are quite satisfactory for the correlation estimator. The Chi-square statistic is computed on a 20 categories division and a null hypothesis of uniformly distributed values may not be rejected at any reasonable level. For the spectral estimator figure 1 shows good performance, but in figure 2 there seems to be some problems with the tails. The use of this signal size indicates that the asymptotic performance of the method is satisfactory for these estimators. Results for some other estimators are discussed in Nordgaard (1990).

I am especially greatful to Docent Urban Hjorth for suggesting this problem and for giving me helpful advice throughout the work. I would also like to thank the Referees for their valuable comments. Computer time on CRAY X/MP-48 has been given by the National Supercomputer Centre in Linköping.

REFERENCES

Efron B. (1979). Bootstrap Methods: Another look at the jackknife. *Ann. Statist.* **7**, 1–26.

Hjorth U. (1987). *Stokastiska Processer—Korrelations- och Spektralteori*. Studentlitteratur, Lund.

Nordgaard A. (1990). On the Resampling of Stochastic Processes using a Bootstrap Approach. Liu-Tek-Lic-1990:23. Linköping, Sweden.

Rosenblatt M. (1985). *Stationary Sequences and Random Fields*. Birkhäuser, Boston.

Applications in Epidemiology and Medical Statistics (8)

EXPLORING HETEROGENEOUS RISK STRUCTURE: COMPARISON OF A BOOTSTRAPPED MODEL SELECTION AND NONPARAMETRIC CLASSIFICATION TECHNIQUE

Peter Dirschedl   *Renate Grohmann*
Dept. of Medical Data Processing, Biometry and Epidemiology   &   Dept. of Psychiatry
Ludwig-Maximilians-Universität München, Marchioninistr.15, D-8000 München 70

Summary
The intention of epidemiological studies is the identification of risk factors. Various techniques are used to assess the quantitative nature of effects, e.g. logistic risk estimation. We display data from an adverse drug reaction study with obvious and strong heterogeneity of risks in strata. Since for this data the origin of heterogeneity is well understood, we compare two methods for their ability of detecting different risk structures. It turns out that the nonparametric classification technique easily uncovers different risk strata, whereas the bootstrapped model selection approach fails. Some insight into the structural properties of bootstrap samples explains the failure of the bootstrap approach.

1. Introduction

From 1979-1986 a collaborative study on *adverse drug reactions ADR* was sponsored by the *Federal Health Agency* (Bundesgesundheitsamt) and has been carried out in German psychiatric hospitals [1]. In the prospective *Intensive Drug Monitoring IDM* branch of the study full information is available on 1107 patients, including complete drug prescription records and the ADR history.
From these patients 395 had been treated with Haloperidol. Our special interest concentrates upon severe side effects like the *extrapyramidal motoric syndrome EPS*. In 182 cases Haloperidol had been imputed for the EPS, and we would like to identify additional risk factors, e.g. the potential influence of diagnosis or various types of psychotropic comedication.

The logistic analysis, applied to the complete data set, resulted in a model including the drugs *Haloperidol*, *Biperiden*, and *tricyclic antidepressants*, together with a very strange interaction term involving *age* and *diagnosis* of patients. The fit was not very good and one can argue, that the risks of comedication are completely different in diagnostic strata, i.e. for patients with *schizophrenia* or *endogeneous depression*.
Actually the logistic analysis *within* this strata led to good fitting models which are understandable and not biased with risk effects from the other stratum. I.e., in group *schizophrenia* the risk factors *Biperiden*, *Perazin* and *Haloperidol* could be identified, the model *(B,P,H)* in short hand, whereas *age* and *tricyclic antidepressants*, *(A,T)*, are the main factors for the group *endogeneous depression*. The estimates for these variables are very stable, significant on the level 0.001 and can be distinguished clearly from other potential factors (all with p>0.2).

In general situations there may exist strata with heterogeneous risks not known *a priori*. Then it would be interesting to know what technique has a fair chance to detect that heterogeneous strata.
We illustrate that the application of a *nonparametric classification technique* can detect risk heterogeneity quite easily. On the other hand, *bootstrapped model selection* procedures have to be used carefully - the bootstrap is not designed for the purpose of detecting strata. A closer inspection of the internal structure of bootstrap samples yields a theoretical explanation of the failure of the bootstrap in the situation considered.

2. Results of the Nonparametric Classification Technique

The nonparametric classification technique (or "CART" approach) is an exploratory method [2]. Its main goal is to construct strata which are internally as similar as possible and externally maximally different with respect to some measure. The procedure starts with all cases and then scans all independent variables, if they support a significant split of the data in two subgroups, thus defining the root of a binary tree. The recursive application of the technique to resulting branches leads to the construction of a classification tree. In the case of a binary dependent variable (e.g. development of syndromes like EPS: *yes / no*) one can show that it is sufficient to split on each level with the variable that reaches a maximum χ^2-value.

Fig. 1a: Nonparametric Classification Analysis (CART-tree)

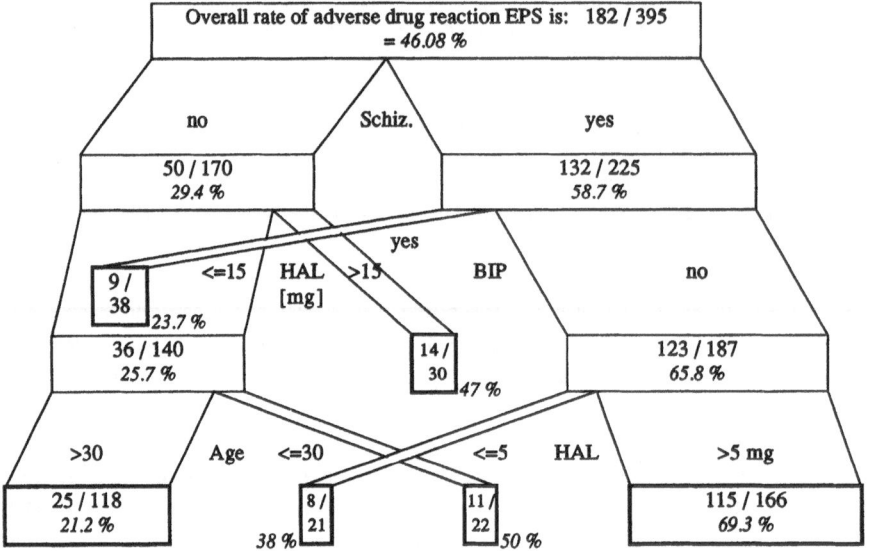

In our data, all independent variables are dichotomous, and the dosis of *Haloperidol, HAL,* being coded as three ordinal classes. At the first split we attain the best result when dividing into the diagnostic groups of *schizophrenia (y/n)*, with a value of $\chi^2 = 33.4$. This partition competes with alternative splits, e.g. by *tricyclic antidepressants (y/n)*, with $\chi^2 = 30.3$, or *Haloperidol* dose with $\chi^2 = 20.2$. Further recursive partitioning leads to the final classification tree of Figure 1a. Note, that after controlling for *diagnosis* the prophylactic usage of *Biperiden (BIP)* shows a clear - and decreasing - effect on EPS rates.

This is one possible look at the data, others may be attained as well. For instance the "look ahead 2" strategy supposed for CART may be applied. This means the following: not only the variable with the "best" split should be used at every step, but the *second best* splits should be tried as well.
In our case the second best competitor to *schizophrenia* is *TAD, tricyclic antidepressants (y/n)*, as noted above, and the final tree resulting from a prior split on *TAD* reveals a risk structure completely different, as shown in Figure 1b. It is quite easy to identify the reason: *TAD* are the drugs of choice for patients with *endogeneous depression*, and therefore the left main branch in Figure 1b mainly consists of those patients (the age effect in this group is well known).

Thus it was possible to identify the heterogeneous risk strata by means of two simple CART analyses, including an application of the "look ahead 2" strategy.

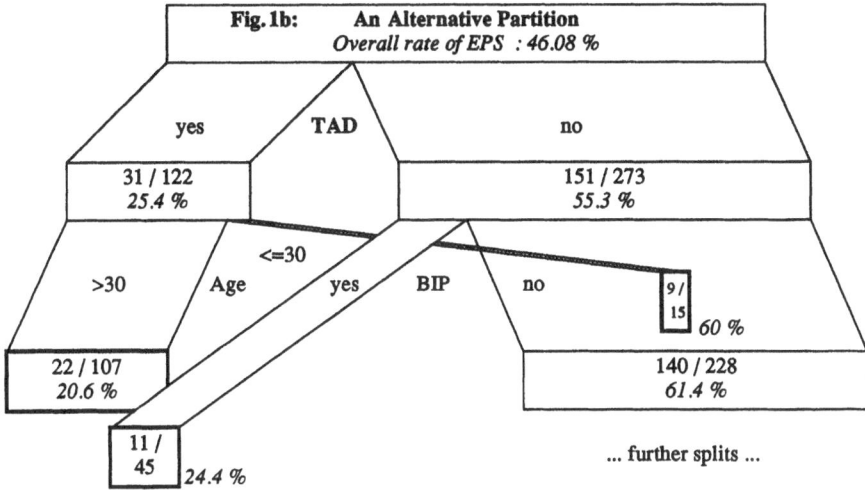

3. Results of the Bootstrapped Model Selection Procedure

Bootstrap techniques for identification of prognostic factors have been proposed for the homogeneous case [3]. Since the parametric bootstrap approach does not make sense in model selection applications we applied the nonparametric bootstrap. We specified the logistic approach, wrote a macro for the *GLIM* package in order to draw the bootstrap samples, to carry out the maximum likelihood estimation and the backward deletion process.

Tab. 1: Frequency of Models Selected after 1,200 Bootstraps

rank	n	ATBPH	%	n	ATBPH	%
		p = 0.001			p = 0.01	
1	227	01101	18.9	384	**11111**	32.0
2	220	**01100**	18.3	248	11101	20.7
3	163	11100	13.6	140	01111	11.7
4	115	10101	9.6	132	01101	11.0
5	99	11101	8.2	89	11100	7.4
6	92	01111	7.7	82	11110	6.8
7	69	01110	5.8	33	10101	2.8
8	68	**11111**	5.7	31	01110	2.6
9	56	11110	4.7	31	**01100**	2.6
10	31	10111	2.6	28	10111	2.3
11	17	00101	1.4	1	11010	0.1

For each trial we started with the full model *(A,T,B,P,H)* - a short hand for the inclusion of *age, tricyclic antidepressants, Biperiden, Perazin* and *Haloperidol* - because the analyses in the original sample stopped with the models *(A,T)* and *(B,P,H)* respectively in the diagnostic strata. For parametric model estimation this is a stringent but necessary and pragmatic restriction to a subset of all possible risk factors.

Altogether 1,200 bootstrap samples were drawn and everytime the backward deletion process was carried out using two different test levels. This means that variables were eliminated if the p-value of the likelihood ratio test exceeded the levels p=0.01 or p=0.001. The computational expense for the bootstrap is considerable, i.e. 30,769 likelihood-ratio tests had to be computed when testing at the level p=0.001. To contrast this impressing number to the expenditure for the CART procedure: the derivation of typical trees as computed here - about 30 independent variables in the "pool", eleven intermediate or final nodes - requires just about 330 χ^2-tests, whose computation is less expensive. The result of the time consuming process, i.e. the models selected most frequently by the bootstrap procedure are listed in Table 1.

Obviously a test level of p=0.01 is much to coarse. In most of the samples the full model *(A,T,B,P,H)* remains accepted, whereas this model has only rank 8 if the much stricter level of 0.001 was applied. So we discuss only the structure obtained at the level 0.001 in the sequel. Note that the *most parsimonious* model *(T,B)* has second rank, only superseded by one of its predecessors in the model hierarchy, which is *(T,B,H)*. The next frequent model is *(A,T,B)*, hierarchically higher as well. In our opinion it does not make sense to derive a final conclusion about importance of variables based on univariate counts of frequencies, or similar counts of marginal frequencies. Since the structure within the models should be retained, and since there exists a natural order of models, we can display the result as a lattice structure, e.g. as sketched in Figure 2 (numbers denote frequency of selected model).

Fig. 2: Lattice of the 10 models selected most frequently

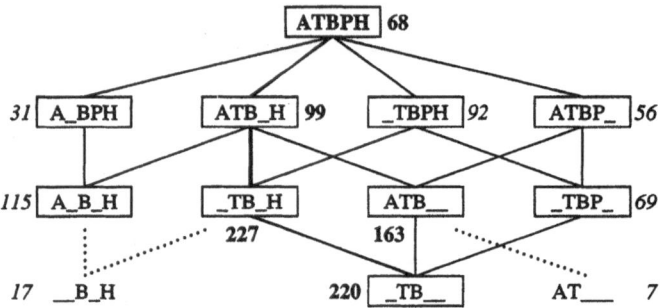

At one glance one can see now that model *(T,B)* is the winner, together with all its direct predecessors. Nearly all paths in the lattice lead to this model and suggest that at least *tricyclic antidepressants (T)* and the prophylactic treatment with *Biperiden (B)* should be considered. Unfortunately, in the original data set these models are not supported at all! Actually there is only one competing path (left side in Figure 2) with models including *Biperiden (B)* and *Haloperidol dose (H)*. This is the only (and weak) indicator in the lattice supporting the model *(B,P,H)* - as originally found in group *schizophrenia*. On the other hand the original model *(A,T)* found for the group *endogeneous depression* is recovered in seven of the trials. Obviously bootstrapped model selection does not work if heterogeneity is present. Therefore one should ask for the reasons.

4. Structural Properties of Bootstrap Samples

It is interesting to look at the structural properties of nonparametric bootstrap samples. The resampling procedure is defined as follows: given a set of n cases { $X_1 , \ldots , X_n$ }, draw a random sample { X_i^* } of size n, independently, with replacement, i.e. $\sim$ *mult* (n, p = 1/n).

Now define an event A_k , if a case is drawn k *times* in n *trials*, with probability p=1/n. It is wellknown that

$$p_k := prob\,(A_k) = \begin{bmatrix} n \\ k \end{bmatrix} p^k\,(1-p)^{n-k}$$

Simple calculations show that $p_{boot} := p_{k>0} = 1 - (1-1/n)^n$, which tends to $1-1/e = 0.632$ as $n \to \infty$. It is important to see that this *selection probability* is fairly *constant* for all finite n of practical interest: e.g. for n=50, $p_{boot} = 0.6358$, for n=2,000, $p_{boot} = 0.6322$, and for the n=323 in this data set, $p_{boot} = 0.6327$. Thus, the expected number of cases selected is $n * p_{boot} = 204$ in this case. Note that - of course - the *sample ratio* of members from the substrata (endogeneous depression, schizophrenia) may vary.

Similar considerations lead to results which may have serious consequences when bootstrapping life table data. For instance, the probability to be drawn *at least twice* is $p_{k>1} = 0.264$, and these multiple cases generate *ties*. Note that the relative ratio of multiple cases *within* the bootstrap sample, $p_{k>1}\,/\,p_{k>0}$, is 0.418 and *constant* again for all relevant n. Hence in expectation there is a fixed and high percentage of ties not neglectable. Unfortunately, until now nothing has been published about the consequences of this intrinsic structural fact, i.e. for the application of bootstrapped model selection procedures on life table data. In our case we want to detect heterogeneity, more precisely to detect heterogeneous risk structures supposed to be present in substrata. Of course a special risk structure may be detected only if the sample drawn via bootstrap is as "clean" as possible. Any contamination with cases from a competing group will bias the estimation in the stratum of interest.

So we might ask for the chance to get a "clean" sample. Pragmatically, we define a sample to be *clean*, if it contains at least 75% of the cases from one single stratum, e.g. from group *schizophrenia*. Since there are 225 cases in that group (and 98 with *endogeneous depression*), the bootstrap sample should contain at least $n_S = 3/4 * 225 = 169$ cases from that group and the remaining n_E from group *endogeneous depression*. Typically, n_{boot} is $n_S + n_E = 204$ in expectation, and the chance of drawing a sample like this is

$$prob = \sum_{i=169}^{204} \begin{bmatrix} 225 \\ i \end{bmatrix} \begin{bmatrix} 98 \\ 204-i \end{bmatrix} / \begin{bmatrix} 323 \\ 204 \end{bmatrix} = 0.000\ 000\ 000\ 024$$

i.e. once in every 41 billion trials, a rare event indeed. It is illustrative to look at Figures 3, where the resulting case mixes are shown after 5,000 (Fig. 3a) and after additional 45,000 (Fig. 3b) bootstraps from 323 cases.

Fig. 3a,b: The Structure of Cases Mixes Selected from Subgroups

n_E = # cases with endogeneous depression

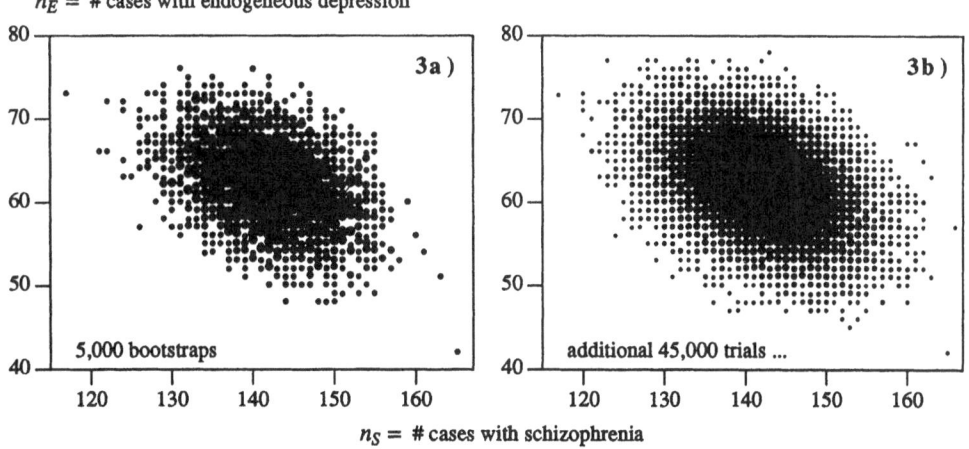

n_S = # cases with schizophrenia

Each point stands for the resulting mix of substrata sizes n_S, n_E within a bootstrap sample (or for some set of samples, because of the discreteness of the data). The first 5,000 samples in Figure 3a show the basic property of the bivariate distribution of (n_S, n_E). After 45,000 additional resamplings all case mixes are represented by only 894 clusters (Fig. 3b), and the point cloud remains concentrated very strongly in comparison to Figure 3a.

A closer look at Figure 3a is given by Figure 4. It displays the first 5,000 bootstraps again, but with some lines added to cut off the regions of *clean* samples desired. Of course, the point cloud lies neatly on the line where substrata sizes n_S and n_E sum up to the 204 cases expected. Dotted lines bound the regions where at least 75% cases from the respective substrata are involved: $n_E \geq 74$ in the upper part (i.e. from group *endogeneous depression*), or $n_S \geq 169$ in the part to the right (i.e. at least 75% from group *schizophrenia*). It can bee seen that the model *(A,T)* related to the *smaller* group *endogeneous depression* may be found more easily than the corresponding model *(B,P,H)* for group *schizophrenia*. Actually, the chance of drawing a "clean" sample from group *endogeneous depression* is about 0.0014. This number compares very well with the resulting lattice as displayed in Figure 2.

Hence, the regions desired typically lie far away from the center of the bootstrap samples attained, and *clean* samples are very rare. Even those samples where at least 153 cases (75% of 204) do come from group *schizophrenia* represent only a small fraction of all trials.

Fig. 4: Desired Regions - and the Reality after 5,000 Bootstraps

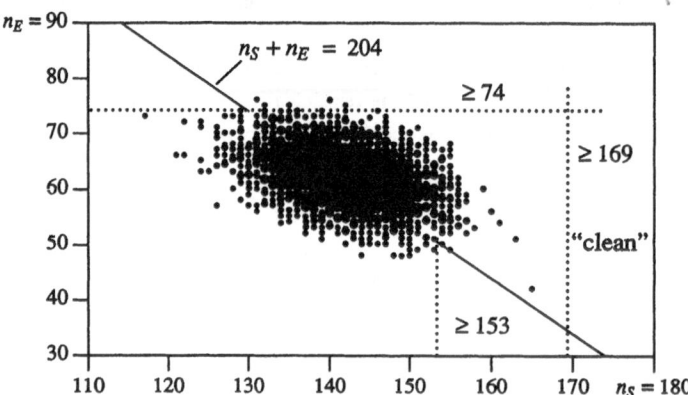

In general, the bootstrap selector is randomly spread over the strata. This is the common situation, yielding the most undesirable case mixes and thus contaminated estimators. If - in the best case - the coverage of a stratum is worth mentioning, this event is rare and only the major stratum can be affected. Even full coverages of the minor stratum are always heavily contaminated. Hence, if there are heterogeneous strata - no matter what number of cases and what risk structures may be present within the groups - the basic structure of nonparametric bootstrap samples lead to the following *conclusions* (the overall situation may be sketched as in Figure 5).

1) In expectation a bootstrap selector always includes a fixed fraction of ~ 63.2% of original cases. This fraction is *constant* for all relevant n, and the relative coverage of substrata determines the outcome of the trial.

2) "Clean" bootstrap samples are very rare, i.e. there is no good chance to obtain a sample that does contain a sufficient and predominant amount of cases (from either group).

3) Because of the heavy contamination the majority of samples drawn will support mixtures of risk structures.

This fact alone explains (partially) the failure of the bootstrap approach - since ML-estimators are very sensitive to contaminated data. Thus the situation is much worse than in the homogeneous case - and even there remains the delicate task to figure out what models are selected more frequently whilst traversing through the complete lattice and being filtered out by the selection procedure.

4) There may be different structural outcomes if different test levels are used. High levels will lead to very coarse structures in the final lattice. So it might be a good advice to carry out the selection process based upon very restrictive p-levels, in order to split up the paths as much as possible.

Fig. 5: The Structural Problem

	schizophrenia	endog. depression	
sample:	n = 225	98	← the strata
general:	(typical selector pattern)		← a typical selector
best case:	204 / 63.3 %		← this coverage is rare!
worst case:	106 / 33 %	98 / 30.3 %	← contamination

5. Discussion

We displayed a data set where considerable reasons suggest the existence of different risk structures in two subgroups. The logistic analysis intended could not detect the different structures - since they can't be expressed conditionally on groups - and failed. The nonparametric classification technique took all variables into consideration and the different risk structures could be revealed easily. Unfortunately this procedure does not lead to risk estimators that can be utilized. On the other hand we illustrated some drawbacks of the bootstrapped model selection approach. First of all, the variables must be preselected - the less the better. Second, the computational expense is immense and the outcome is doubtful. Apart from other problems, i.e. the dependence on the test level used and difficulties to identify a risk structure in the lattice of models, there remains a much more striking fact: The basic structural property of a nonparametric bootstrap leads to samples which are typically contaminated and can not reveal heterogeneity at all.

Acknowledgements

The authors would like to thank the referees for their very helpful comments.

References

[1] THE AMÜP STUDY GROUP (1989). *Final Report on Drug Surveillance in Psychiatry: Intensive Drug Monitoring*. Preprint (available from the authors upon request - in German)

[2] BREIMAN, L.; FRIEDMAN, J.H.; OLSHEN, R.A., STONE, C.J. (1984). *Classification and Regression Trees*. Monterey, Wadsworth.

[3] CHEN, Ch.; GEORGE, S.L. (1985). The Bootstrap and Identification of Prognostic Factors via Cox's Proportional Hazards Regression Model. *Statistics in Medicine* 4, 39-46.

BOOTSTRAPPING CURRENT LIFE TABLE ESTIMATORS

Amanda L. Golbeck, Ph.D.
Department of Statistics, University of Oxford
1 South Parks Road, Oxford OX1 3TG, United Kingdom

Abstract. This research involves development of appropriate bootstrap methods for current life table estimators, where the data have a structure that is more complex than that arising from independent and identically distributed sampling. In general outline, the bootstrap approach described here involves conceptualization of a hypothetical cohort to correspond to each row of the life table. Simple random samples with replacement are repeatedly taken from the "observations" for each hypothetical cohort. These contribute to the construction of a set of simulated life tables which provides approximate sampling distributions for the estimators of the biometric functions contained within it. Methods are illustrated using data for the 1980 San Diego County (U.S.A.) population aged 65 and over.

Efron (1981) and Reid (1981) each developed bootstrap resampling plans for randomly censored data summarized by the Kaplan-Meier estimate. Akritas (1986) went on to study the asymptotic behavior of the bootstrapped Kaplan-Meier estimator with both resampling plans, and concluded that only Efron's bootstrap method for randomly censored data can be used to construct asymptotically correct confidence bands.

The present study deals with bootstrapping another class of survival estimators, namely current life table estimators. The life table in its modern form is a model of survival that can be dated back to Graunt (1662) and Halley (1693). Current life tables involve construction of hypothetical cohorts that experience mortality estimated from current study population data. Such tables give cross-sectional summaries of mortality during a current year. Methods of constructing current life tables generally involve computing an estimate of the probability of death in an age interval from the corresponding age-specific death rate (this rate being calculated as the number of deaths during the current year to individuals in that age group divided by the midyear population size in that age group), after which all of the other quantities in the life table can be readily calculated. Current life tables are thus dependent entirely upon the age-specific death rates from the current study population.

Here we a) develop a bootstrap resampling plan appropriate for current population data summarized by life table estimators, b) empirically demonstrate these methods to suggest support for Chiang's (1960b) formulas for estimating variances of current life table estimators, and c) show how these methods can be used to produce variance estimates that will generally be smaller than Chiang's. The presentation focuses on basic ideas.

We observe $\{y_j\}_{j=1}^{D}$, the exact ages at death of the individuals in the study population who died during the study year (determined from vital statistics certificates of death). We confine our discussion to variance estimation for the following two biometric functions in the life table computed from these observations: $\hat{q}_i$, the estimated conditional probability of death within the i-th age interval, and $\hat{p}_{0i}$, the estimated probability of surviving from age x_0 to age x_i, $i=0,1,...,\omega$.

Define D_i as the number of y_j's that fall into the i-th age interval in the life table. Golbeck (1986) describes alternative estimators for $\hat{q}_i$: Here we let $\hat{q}_i = \frac{n_i D_i}{P_i+(1-a_i)n_i D_i}$, where n_i is the width of the i-th age interval, P_i is the midyear population size in the i-th age interval (determined from census data), and a_i is the average fraction of the i-th age interval lived by those who die in the interval (determined from the y_j's or else taken to equal .5). Also let $\hat{p}_{0i} = \hat{p}_0 \hat{p}_1...\hat{p}_{i-1}$, where $\hat{p}_h=1-\hat{q}_h$.

For variance estimation for $\hat{q}_i$, Chiang writes $\hat{q}_i = \frac{D_i}{N_i}$, where N_i is the size of a hypothetical cohort alive at the beginning of the i-th age interval, if the force of mortality acts on each of the N_i individuals. Then he writes the estimated variance of $\hat{q}_i$ as $S^2(\hat{q}_i) = \frac{q_i(1-q_i)}{N_i}$. However, N_i is unknown in current life table applications. Looking at $\hat{q}_i = \frac{n_i D_i}{P_i+(1-a_i)n_i D_i}$, Chiang suggests taking $N_i = \frac{P_i}{n_i}+(1-a_i)D_i$, treating N_i as constant. For variance estimation for $\hat{p}_{0i}$, Chiang uses the delta method with $\hat{p}_{0i} = \hat{p}_0 \hat{p}_1...\hat{p}_{i-1}$ to write the estimated variance of $\hat{p}_{0i}$ as a function of the variance of the $\hat{p}_h$'s as follows: $S^2(\hat{p}_{0i}) = \hat{p}_{0i}^2 \sum_{h=0}^{i-1} \hat{p}_h^{-2} S^2(\hat{p}_h)$.

To motivate the need for formulation of resampling plans for current population data summarized by life table estimates, consider a naive bootstrap method which assumes that $\{y_j\}_{j=1}^{D}$ are independent and identically distributed with distribution function $\hat{F}$. Taking a bootstrap sample $\{y_j^*\}_{j=1}^{D}$ involves taking a simple random sample with replacement from $\hat{F}$. Define D_i^* as the number of y_j^*'s that fall into the i-th age interval. To correspond to Chiang's treatment, let $\hat{q}_i^* = \frac{D_i^*}{N_i} = \frac{n_i D_i^*}{P_i+(1-a_i)n_i D_i}$. Also let $\hat{p}_{0i}^* = \hat{p}_0^* \hat{p}_1^*...\hat{p}_{i-1}^*$. The naive bootstrap involves drawing B samples of size D from $\hat{F}$, computing $\hat{q}_i^*$ and $\hat{p}_{0i}^*$ a total of B times, and computing the sample variance (see Efron, 1979a, and 1979b) of $\hat{q}_i^*$ and $\hat{p}_{0i}^*$ under this sampling. The vector $D^* = (D_0^*, D_1^*, ..., D_\omega^*)$ in the naive bootstrap is a $(\omega+1)$-dimensional multinomial with sample size D and vector of probabilities $(\frac{D_0}{D}, \frac{D_1}{D}, ..., \frac{D_\omega}{D})$. The variance of D_i^* is $V(D_i^*) = D_i(1-\frac{D_i}{D})$. The covariance of D_i^* and D_j^* is $\sigma(D_i^*, D_j^*) = -\frac{D_i D_j}{D}$, for $i \neq j$.

The multinomial nature of the naive bootstrap is not compatible with existing statistical theory of the life table. The variance of $\hat{q}_i^*$ under the naive bootstrap method is $V(\hat{q}_i^*) = (\frac{1}{N_i^2})D_i(1-\frac{D_i}{D})$. Chiang's variance estimation formula may be rewritten as $S^2(\hat{q}_i) = (\frac{1}{N_i^2})D_i(1-\frac{D_i}{N_i})$ to highlight its difference from $V(\hat{q}_i^*)$. The covariance of $\hat{p}_i^*$ and $\hat{p}_j^*$ for $i \neq j$ under the naive bootstrap method is $\sigma(\hat{p}_i^*, \hat{p}_j^*) = \frac{1}{N_i}\frac{1}{N_j}(-\frac{D_i D_j}{D})$. Note that in virtually all life table applications the D_h's are nonzero, and hence $\sigma(\hat{p}_i^*, \hat{p}_j^*) \neq 0$ for $i \neq j$. For estimation of the variance of $\hat{p}_{0i}$, Chiang uses the theoretical result that $\sigma(\hat{p}_i, \hat{p}_j)=0$ for $i \neq j$ (Chiang 1960a).

These incompatibilities may be eliminated by associating the y_j's with the hypothetical cohorts alive at the beginning of each age interval. For the i-th hypothetical cohort or stratum, we "observe" $\{z_{i,k}\}_{k=1}^{N_i}$: The first D_i of these are arbitrarily set equal to the y_j's that fall into the i-th age interval; the remaining N_i-D_i are set equal to zero to mark individuals in the hypothetical cohort who have not died during the study year. The $z_{i,k}$'s within stratum i are independent and identically distributed observations from a distribution F_i. For $i \neq i'$, $z_{i,k}$ and $z_{i',k'}$ are independent but not necessarily identically distributed.

An obvious version of the bootstrap for these hypothetical cohort data involves taking a simple random sample $\{z_{i,k}^*\}_{k=1}^{N_i}$ with replacement from the sample $\{z_{i,k}\}_{k=1}^{N_i}$ in the i-th stratum, independently for each stratum; calculating D_i^* as the number of $z_{i,k}^*$'s that are nonzero and fall into the i-th age interval; calculating $\hat{q}_i^*$ as discussed below; and calculating $\hat{p}_{0i}^* = \hat{p}_0^* \hat{p}_1^* \ldots \hat{p}_{i-1}^*$. This procedure is repeated independently B times to obtain a sample variance of $\hat{q}_i^*$ and $\hat{p}_{0i}^*$. D_i^* in the hypothetical cohort bootstrap is binomial with sample size N_i and probability $\frac{D_i}{N_i}$. The variance of D_i^* is $V(D_i^*) = D_i(1-\frac{D_i}{N_i})$, and the covariance of D_i^* and D_j^* is $\sigma(D_i^*, D_j^*) = 0$ for $i \neq j$. These features are compatible with standard life table theory.

We will consider three cases for $\hat{q}_i^*$. I: $\hat{q}_i^* = \frac{D_i^*}{N_i} = \frac{n_i D_i^*}{P_i + (1-a_i)n_i D_i}$. The bootstrap variance estimate $V^*(\hat{q}_i)$ in this case corresponds to Chiang's $S^2(\hat{q}_i)$. II: $\hat{q}_i^* = \frac{n_i D_i^*}{P_i + (1-a_i)n_i D_i^*}$. This corresponds to $S^2(\hat{q}_i) \approx (\frac{P_i}{n_i N_i^2})^2 D_i(1-\frac{D_i}{N_i})$, obtained by applying the delta method with $\hat{q}_i = \frac{n_i D_i}{P_i + (1-a_i)n_i D_i}$. $S^2(\hat{q}_i)$ for case II is equal to the corresponding quantity for case I multiplied by the factor $(\frac{P_i}{n_i N_i})^2$. Since this factor is always less than 1, the variance estimates for case II will always be smaller than those for case I. Chiang suggested this type of approach for estimating the variance of the age-specific death rate, in order to include additional variability (Chiang, 1986). III: $\hat{q}_i^* = \frac{n_i D_i^*}{P_i + (1-a_i^*)n_i D_i^*}$. This is a generalization of case II to include the variability of a_i. Since the variance estimates for case II will always be smaller by the factor than those for case I, and since we expect the contribution of a_i to the total variation to be small, we therefore expect the variance estimates for case III to usually be smaller than those for case I.

The study population for the illustration in this paper consists of San Diego County residents age 65 and over in 1980. Two sources of data were required: Computerized unit-record vital statistics certificates of death from the San Diego County Department of Health were used to calculate the D_i's and a_i's; and 1980 United States Census of Population and Housing reference manuals were used to determine the P_i's. We used a CRAY X-MP/48 supercomputer and the Wichmann and Hill pseudo-random number generator (Griffiths and Hill, 1985) to resample B=10,000 times from the set of D=9,073 deaths to members of the study population. Using $N_i = \frac{P_i}{n_i} + (1-a_i)D_i$, we have for this population N_0=14,194; N_1=11,174; N_2=8,070; and N_3=5,176. Thus, 10,000 simulated life tables were constructed which, taken as a set, provided the summary data compiled into the Table shown below.

Table: Selected Current Life Table Results
for San Diego County Residents Ages 65 and Over in 1980
(based upon D=9,073 deaths, and B=10,000 bootstrap trials)

(a) Results for $\hat{q}_i$, the estimated conditional probability of death within the i-th age interval

			Variance Estimates for $\hat{q}_i$ Multiplied by 10^6				
			(a)	(b)	(c)	(d)	(e)
			Chiang	Bootstrap		Bootstrap	Bootstrap
i	x_i-x_{i+1}	$\hat{q}_i$	(1960b)	Case I	Delta Method	Case II	Case III
0	65-70	0.1041	6.5720	6.5089	5.9292	5.8724	5.8764
1	70-75	0.1482	11.2969	11.0651	9.6830	9.4844	9.4938
2	75-80	0.2174	21.0796	20.5014	16.7660	16.3073	16.4421
3	80-85	0.3232	42.2641	43.0106	29.5872	30.1178	50.5633
4	85+	1.0000	0.0000	0.0000	0.0000	0.0000	0.0000

(b) Corresponding results for $\hat{p}_{0i}$, the estimated probability of surviving from age x_0 to age x_i

			Variance Estimates for $\hat{p}_{0i}$ Multiplied by 10^6				
			(a)	(b)	(c)	(d)	(e)
			Chiang	Bootstrap		Bootstrap	Bootstrap
i	x_i-x_{i+1}	$\hat{p}_{0i}$	(1960b)	Case I	Delta Method	Case II	Case III
0	65-70	1.0000	0.0000	0.0000	0.0000	0.0000	0.0000
1	70-75	0.8959	6.5720	6.5092	5.9292	5.8718	5.8765
2	75-80	0.7631	13.8354	13.5566	12.0737	11.8317	11.8500
3	80-85	0.5972	20.7503	20.1878	17.1592	16.6961	16.8105
4	85+	0.4042	24.5798	24.4182	18.4131	18.2647	18.4403

The variance estimates shown in the Table by formula and by hypothetical cohort bootstrap method are close for all age intervals under both cases I and II (comparing for case I column (a) with (b); comparing for case II column (c) with (d)). The variance estimates in column (e) were produced using the hypothetical cohort bootstrap method under case III. These may be compared to Chiang's variance estimates from column (a) to see that the bootstrap estimates are generally smaller, as expected.

We conclude for the current life table problem that the hypothetical cohort bootstrap method has promise; that Chiang's variance estimation formulas are supported; and that bootstrap methods can be used to produce variance estimates that incorporate all of the variability attributable to the y_j's. We are currently investigating Poisson, as alternative to binomial, assumptions.

References

AKRITAS, M. G. (1986), "Bootstrapping the Kaplan-Meier Estimator," *Journal of the American Statistical Association,* 81, 1032-1038.

CHIANG, C. L. (1960a), "A Stochastic Study of the Life Table and Its Applications: I. Probability Distributions of the Biometric Functions," *Biometrics,* 16, 618-635.

CHIANG, C. L. (1960b), "A Stochastic Study of the Life Table and Its Applications: II. Sample Variance of the Observed Expectation of Life and Other Biometric Functions," *Human Biology,* 32, 221-238.

CHIANG, C. L. (1986), "Discussion on the Paper by David R. Brillinger: The Natural Variability of Vital Rates and Associated Statistics," *Biometrics,* 42, 729-732.

EFRON, B. (1979a), "Bootstrap Methods: Another Look at the Jackknife," *Annals of Statistics,* 7, 1-26.

EFRON, B. (1979b), "Computers and the Theory of Statistics: Thinking the Unthinkable," *SIAM Review,* 21, 460-480.

EFRON, B. (1981), "Censored Data and the Bootstrap," *Journal of the American Statistical Association,* 76, 312-319.

GOLBECK, A. L. (1986), "Probabilistic Approaches to Current Life Table Estimation," *The American Statistician,* 40, 185-190.

GRAUNT, J. (1662), *Natural and Political Observations Made upon the Bills of Mortality.* Reprinted by the Johns Hopkins Press, Baltimore, 1939.

GRIFFITHS, P. AND HILL, I.D., eds. (1985), *Applied Statistics Algorithms,* Chichester, United Kingdom: Ellis Horwood Limited.

HALLEY, E. (1693), "An Estimate of the Degrees of the Mortality of Mankind, Drawn from Curious Tables of the Births and Funerals at the City of Breslau," *Philosophical Transactions of the Royal Society of London,* 17, 596-610.

REID, N. (1981), "Estimating the Median Survival Time," *Biometrika,* 68, 601-608.

JACKKNIFING ESTIMATORS OF A COMMON ODDS RATIO
FROM SEVERAL 2x2 TABLES

Iris Pigeot
Universität Dortmund, Fachbereich Statistik
Postfach 50 05 00, D-4600 Dortmund 50

Summary
This paper gives a review of certain jackknife estimators of a common odds ratio in the situation of several 2x2 tables. These jackknife estimators are based on the classical Mantel-Haenszel estimator and on some asymptotically efficient noniterative estimators. Finite-sample results and asymptotic properties for increasing sample sizes, but a fixed number of tables are summarized. Some open problems in this field of research are indicated.

1. Introduction

Consider K independent pairs of independent binomial random variables X_{0k} and X_{1k} with parameters N_{0k}, p_{0k} and N_{1k}, p_{1k}, respectively, k=1,...,K, $1 < K < \infty$. The observations of these random variables can be shown in K 2x2 contingency tables. Let each individual odds ratio ψ_k be defined as $\{p_{1k}(1-p_{0k})\}/\{p_{0k}(1-p_{1k})\}$, k=1,...,K. Assume that all ψ_k, k=1,...,K, are finite and equal to a common odds ratio ψ. Under this assumption of homogeneity, many estimators of ψ have been suggested in the literature (see e.g. Woolf, 1955; Haldane, 1955; Mantel, Haenszel, 1959; Breslow, Liang, 1982; Davis, 1985; Gastwirth, Greenhouse, 1987; Pigeot, 1989, 1990a, 1990b). Several of these proposals are reviewed for instance in Hauck (1989).

For the case of increasing sample sizes, but a fixed number of tables and for the alternative case of an increasing number of tables, the asymptotic properties of such odds ratio estimators have been discussed extensively in the literature (see eg. Gart, 1962; Hauck, 1979; Breslow, 1981; Tarone et al, 1983; Guilbaud, 1983, Davis, 1985; Pigeot, 1989, 1990a). Besides these investigations in which it turned out that most estimators have desirable asymptotic properties such as consistency and asymptotic normality, simulation studies have been carried out to get an idea of the finite-sample behaviour of odds ratio estimators (e.g. McKinlay, 1975, 1978; Hauck et al, 1982; O'Gorman et al, 1988; Pigeot, 1990b, c). It turned out for instance that the simply constructed and well-known Mantel-Haenszel estimator (Mantel, Haenszel, 1959) shows good results even in comparison with the asymptotically efficient, but iterative Maximum-Likelihood estimator which is often hard to calculate. The noniterative Mantel-Haenszel estimator seems also to be preferable to the Woolf estimator (Woolf, 1955) with respect to the bias. But all odds ratio estimators suffer from being biased which may be serious with small sample sizes. Thus, additional efforts have been made to find suitable methods for a bias reduction. One possibility consists in adding a constant t to each cell entry, e.g. t=0.5 or t=0.25 (c.f. Haldane, 1955; Anscombe, 1956; Hitchcock, 1962; Gart, Zweifel, 1967) before calculating so called 'modified' estimators based e.g. on the Woolf estimator. These modified versions of the Mantel-Haenszel and of the Woolf estimator have also been compared in the simulation study by Hauck et al (1982). Another idea to reduce the bias is based on resampling techniques such as the jackknife principle which will be considered here.

In the analysis of categorical data the jackknife principle has already been applied by several authors. For instance, Fleiss and Davies (1982) jackknifed functions of multinomial frequencies to get a useful method for bias reduction, but also for variance estimation. With this latter goal in mind, Breslow and Liang (1982) proposed a jackknife estimator of the common log odds ratio based on the logarithm of the Mantel-Haenszel estimator. That means, Breslow and Liang were mainly interested in the corresponding jackknife variance as an estimator of the Mantel-Haenszel variance. The type of jackknifing which was suggested in the paper by Breslow and Liang will also be discussed here in a more general presentation. In addition, Parr and Tolley (1982) considered functions of multinomial frequencies and discussed asymptotic properties of resulting jackknife estimators. In this case, asymptotic normality could be shown by using general results for jackknifed U-statistics derived by Arvesen (1969). In the case of a matched-pair design, jackknifed point estimators of the odds ratio were discussed by Jewell (1984).

In the following, two different approaches for constructing jackknife estimators of a common odds ratio are presented. Both approaches are based on omitting one observation when calculating the pseudo-values. But they differ in what is perceived as 'one observation'.

First each 2x2 table is considered as one observation. This approach was applied by Breslow and Liang (1982) to the logarithm of the Mantel-Haenszel estimator. Jackknife estimators of this type, in the following referred to as of type I, are appropriate when dealing with a large number of contingency tables. Thus, a jackknife estimator of type I based on any estimator $\hat{\psi}$ of the common odds ratio ψ is defined as the following arithmetic mean

$$J^I := \frac{1}{K} \sum_{k=1}^{K} J_k^I \text{ with } J_k^I := K\hat{\psi} - (K-1)\hat{\psi}_k^I, \tag{1.1}$$

where $\hat{\psi}_k^I$ is of the same type as $\hat{\psi}$, but it is calculated after omitting the k-th table, k=1,...,K. The jackknife variance $V(J^I)$ results in

$$V(J^I) = \frac{1}{K(K-1)} \sum_{k=1}^{K} \left(J_k^I - J^I \right)^2.$$

The second approach is based on the fact that each binomial variable X_{jk} is a sum of independent Bernoulli trials with parameter p_{jk}, j=0,1, k=1,...,K. Thus, the outcome of each Bernoulli trial can be treated as one observation to be omitted when calculating the pseudo-values. When jackknifing an estimator $\hat{\psi}$ of ψ in this way it should be taken into account that although there are N pseudo-values there are only 4K distinct values $J_k^a,...,J_k^d$ with $J_k^a := N\hat{\psi} - (N-1)\hat{\psi}_k^a$, J_k^b, J_k^c and J_k^d, k=1,...,K, respectively. Here $\hat{\psi}_k^a,...,\hat{\psi}_k^d$ are of the same type as $\hat{\psi}$, but they are calculated after reducing the number of observations in the k-th table and in the appropriate cells by one. The cells are usually marked by a,...,d.

Each of these distinct values exists exactly X_{1k}, $N_{1k}-X_{1k}$, X_{0k} and $N_{0k}-X_{0k}$-times, respectively. Hence this jackknife estimator, referred to as of type II, which is defined as the arithmetic mean of the N pseudo-values, based on an estimator $\hat{\psi}$ reads as follows (Pigeot, 1989)

$$J^{II} = N\hat{\psi} - \frac{N-1}{N} \sum_{k=1}^{K} \left(X_{1k}\hat{\psi}_k^a + (N_{1k}-X_{1k})\hat{\psi}_k^b + X_{0k}\hat{\psi}_k^c + (N_{0k}-X_{0k})\hat{\psi}_k^d \right). \tag{1.2}$$

The jackknife variance of this estimator $V(J^{II})$ can be calculated as

$$V(J^{II}) = \frac{N-1}{N} \sum_{k=1}^{K} \left(X_{1k}\hat{\psi}_k^{a2} + (N_{1k}-X_{1k})\hat{\psi}_k^{b2} + X_{0k}\hat{\psi}_k^{c2} + (N_{0k}-X_{0k})\hat{\psi}_k^{d2} \right) - \frac{1}{N-1} \left(J^{II} - N\hat{\psi} \right)^2.$$

These two approaches are now applied to the classical Mantel-Haenszel estimator

$$\hat{\psi}_{MH} = \frac{\sum\limits_{k=1}^{K} \hat{p}_{1k}\hat{q}_{0k} \frac{N_{1k}N_{0k}}{N_k}}{\sum\limits_{k=1}^{K} \hat{p}_{0k}\hat{q}_{1k} \frac{N_{1k}N_{0k}}{N_k}}, \tag{1.3}$$

with $N_k := N_{0k} + N_{1k}$, $\hat{p}_{jk} := X_{jk}/N_{jk}$, $\hat{q}_{jk} := 1-\hat{p}_{jk}$, j=0,1, k=1,...,K, and to a class of asymptotically efficient noniterative estimators (Pigeot, 1990a)

$$\hat{\psi}_g = g\left(\sum_{k=1}^{K} \hat{w}_k g^{-1}(\hat{\psi}_k)/ \hat{w} \right); \, g \in G, \tag{1.4}$$

with G consisting of all nonnegative and real valued functions which are strictly monotone and differentiable and

$$\hat{w}_k^{-1} := \left(N_{1k}\hat{p}_{1k}\hat{q}_{1k} \right)^{-1} + \left(N_{0k}\hat{p}_{0k}\hat{q}_{0k} \right)^{-1}, \, \hat{w} := \sum_{k=1}^{K} \hat{w}_k, \, \hat{\psi}_k := \{\hat{p}_{1k}\hat{q}_{0k}\}/\{\hat{p}_{0k}\hat{q}_{1k}\}, \, k=1,...,K.$$

This class of estimators contains e.g. the well-known Woolf estimator $\hat{\psi}_{exp} =: \hat{\psi}_W$ (Woolf, 1955) which can easily be seen by choosing g as the exponential function.

It is the purpose of this paper to survey some results for jackknife estimators of the odds ratio. These results are partially discussed in detail in other papers by the author. Therefore detailed proofs will not be given here. It may, however, be mentioned that the proofs concerning the asymptotic properties are based on tools such as a generalization of Slutsky's Theorem (c.f. Serfling, 1980, p. 24) and stochastic expansions of the estimators.

2. Asymptotic properties

We consider the case that the sample sizes N_{0k} and N_{1k}, k=1,...,k, within each table tend to infinity while the number of tables remains fixed. As it is explicitly described below, asymptotic normality can be shown under the condition:

$$N_{jk} = \lambda_{jk}N, \; N = \sum_{k=1}^{K} N_k, \; 0 < \lambda_{jk} < 1, \; j=0,1, \; \sum_{k=1}^{K} \lambda_k = 1, \; \lambda_k := \lambda_{1k}+\lambda_{0k}, \; k=1,...,K, \qquad (2.1)$$

and consistency of the jackknife estimators can be proved under the slightly weaker condition

$$N_{jk} = \lambda_{jk}N + o(N), \; N \to \infty, \; N = \sum_{k=1}^{K} N_k, \; 0 < \lambda_{jk} < 1, \; j=0,1, \; k=1,...,K, \; \sum_{k=1}^{K} \lambda_k = 1. \qquad (2.1')$$

Both conditions were also assumed by other authors dealing with the asymptotic behaviour of estimators of a common odds ratio for increasing sample sizes, but a fixed number of strata (c.f. Gart, 1962; Hauck, 1979; Nurminen, 1981).

Jackknife estimator of type I based on the Mantel-Haenszel estimator

As already mentioned, Breslow and Liang (1982) jackknifed the logarithm of the Mantel-Haenszel estimator $\hat{\psi}_{MH}$ by applying the first approach described above. An exponential transformation yields the following estimator of ψ, referred to as Breslow-Liang estimator $\hat{\psi}_{BL}$, where the term in parenthesis is calculated according to (1.1):

$$\hat{\psi}_{BL} := \exp \left(K \log \hat{\psi}_{MH} - \frac{K-1}{K} \sum_{k=1}^{K} \log \hat{\psi}_{MH,k}^{I} \right) \text{ with } \hat{\psi}_{MH,k}^{I} := \frac{\displaystyle\sum_{i=1,i\neq k}^{K} \hat{p}_{1i}\hat{q}_{0i} \frac{N_{1i}N_{0i}}{N_i}}{\displaystyle\sum_{i=1,i\neq k}^{K} \hat{p}_{0i}\hat{q}_{1i} \frac{N_{1i}N_{0i}}{N_i}}, \; k=1,...,K.$$

Under assumption (2.1'), $\hat{\psi}_{BL}$ is a consistent estimator of ψ (c.f. Pigeot, 1989). And, assuming (2.1), it can be shown, that

$$\left(\hat{\psi}_{BL} - \psi \right) \text{Var}_A(\hat{\psi}_{BL})^{-1/2}$$

is asymptotically standard normally distributed, where the asymptotic variance is given as

$$\text{Var}_A(\hat{\psi}_{BL}) = \psi^2 \sum_{k=1}^{K} \left\{ \frac{K}{C} - \frac{K-1}{K} \sum_{i=1,i\neq k}^{K} B_i^{-1} \right\}^2 v_k^2 w_k^{-1},$$

$$C := \sum_{l=1}^{K} v_l, \; B_i := \sum_{l=1,l\neq i}^{K} v_l, \; w_k^{-1} := \left(N_{1k}p_{1k}q_{1k} \right)^{-1} + \left(N_{0k}p_{0k}q_{0k} \right)^{-1}, \; w := \sum_{k=1}^{K} w_k, \; v_k := p_{0k}q_{1k}N_{1k}N_{0k}/N_k,$$
$$q_{jk} = 1 - p_{jk}, \; j=0,1, \; k=1,...,K.$$

In general the Breslow-Liang estimator is asymptotically inefficient. In the degenerate case, however, where there is no need to stratify, i.e. $p_{jk}=p_j$, $j=0,1$, $N_{1k}/N_k = \rho$, $\rho > 0$, $N_k=N/K$, $k=1,...,K$, $\hat{\psi}_{BL}$ is asymptotically efficient. This can be seen by comparing its asymptotic variance with that of the Mantel-Haenszel estimator, the asymptotic variance of which is identical with that of the asymptotically efficient maximum likelihood estimator of ψ in this degenerate case (Gart, 1962; Tarone et al, 1983).

Jackknife estimators of type I based on some asymptotically efficient noniterative estimators

According to the construction of the above transformed jackknife estimator $\hat{\psi}_{BL}$, the application of the first jackknife approach to the noniterative estimators $\hat{\psi}_g$, $g \in G$, yields the following estimators J^I_g, $g \in G$:

$$J^I_g = Kg^{-1}(\hat{\psi}_g) - \frac{K-1}{K} \sum_{k=1}^{K} g^{-1}(\hat{\psi}^I_{g,k}) \quad \text{with} \quad g^{-1}(\hat{\psi}^I_{g,k}) := \frac{\sum\limits_{i=1,i\neq k}^{K} \hat{w}_i g^{-1}(\hat{\psi}_i)}{\sum\limits_{i=1,i\neq k}^{K} \hat{w}_i} , \quad k=1,...,K.$$

Under the corresponding assumptions, the transformed estimators $g(J^I_g)$ are also consistent and asymptotically normal with asymptotic expansion ψ and asymptotic variance

$$\text{Var}_A(g(J^I_g)) = \psi^2 \sum_{k=1}^{K} w_k \left(\frac{K}{w} - \frac{K-1}{K} \sum_{i=1,i\neq k}^{K} \left(\sum_{l=1,l\neq i}^{K} w_l \right)^{-1} \right)^2 \quad \text{(Pigeot, 1989)}.$$

Except from the degenerate case described above, the estimators $g(J^I_g)$ are in general asymptotically inefficient, too. Moreover it can be seen, that for $\psi_k = \psi = 1$ the asymptotic variance of $g(J^I_g)$ is identical with that of $\hat{\psi}_{BL}$.

Jackknife estimator of type II based on the Mantel-Haenszel estimator

According to (1.2), the jackknife estimator of type II based on $\hat{\psi}_{MH}$ reads as follows:

$$J^{II}_{MH} = N\hat{\psi}_{MH} - \frac{N-1}{N} \sum_{k=1}^{K} \left(X_{1k}\hat{\psi}^a_{MH,k} + (N_{1k}-X_{1k})\hat{\psi}^b_{MH,k} + X_{0k}\hat{\psi}^c_{MH,k} + (N_{0k}-X_{0k})\hat{\psi}^d_{MH,k} \right), \text{ where}$$

$$\hat{\psi}^a_{MH,k} := \frac{(X_{1k}-1)(N_{0k}-X_{0k})\frac{1}{N_k-1} + \sum\limits_{i=1,i\neq k}^{K} \hat{p}_{1i}\hat{q}_{0i}\frac{N_{1i}N_{0i}}{N_i}}{(N_{1k}-X_{1k})X_{0k}\frac{1}{N_k-1} + \sum\limits_{i=1,i\neq k}^{K} \hat{q}_{1i}\hat{p}_{0i}\frac{N_{1i}N_{0i}}{N_i}}$$

and $\hat{\psi}^b_{MH,k}$, $\hat{\psi}^c_{MH,k}$ and $\hat{\psi}^d_{MH,k}$, $k=1,...,K$, are defined analogously. After some analytical manipulations, J^{II}_{MH} turns out to be consistent under assumption (2.1'). Moreover, one can show the asymptotic equivalence of $N^{1/2}J^{II}_{MH}$ and $N^{1/2}\hat{\psi}_{MH}$ (Pigeot, 1989). Since the Mantel-Haenszel estimator is asymptotically normal under (2.1) (c.f. Hauck, 1979; Guilbaud, 1983), asymptotic normality of J^{II}_{MH} with asymptotic expectation ψ and the same asymptotic variance as $\hat{\psi}_{MH}$ follows by the Theorem of Cramér (c.f. Gänssler, Stute, 1977, p. 68). Because of the identity of the asymptotic variances of J^{II}_{MH} and $\hat{\psi}_{MH}$ the conditions for the asymptotic efficiency of J^{II}_{MH} are the same as those for the efficiency of

$\hat{\psi}_{MH}$ derived by Tarone, Gart and Hauck (1983).

Jackknife estimators of type II based on some asymptotically efficient noniterative estimators

Similarly to J_{MH}^{II} the jackknife estimators of type II based on the estimators $\hat{\psi}_g$, $g \in G$, can be derived as

$$J_g^{II} = N g^{-1}(\hat{\psi}_g) - \frac{N-1}{N} \sum_{k=1}^{K} \left(X_{1k} g^{-1}(\hat{\psi}_{g,k}^a) + (N_{1k}-X_{1k}) g^{-1}(\hat{\psi}_{g,k}^b) + X_{0k} g^{-1}(\hat{\psi}_{g,k}^c) + (N_{0k}-X_{0k}) g^{-1}(\hat{\psi}_{g,k}^d) \right), \ g \in G,$$

where $g^{-1}(\hat{\psi}_{g,k}^a) := \dfrac{\left(\dfrac{N_{1k}-1}{(X_{1k}-1)(N_{1k}-X_{1k})} + \dfrac{N_{0k}}{X_{0k}(N_{0k}-X_{0k})} \right)^{-1} g^{-1}\left(\dfrac{(X_{1k}-1)(N_{0k}-X_{0k})}{X_{0k}(N_{1k}-X_{1k})} \right) + \sum\limits_{i=1,i\neq k}^{K} \hat{w}_i g^{-1}(\hat{\psi}_i)}{\left(\dfrac{N_{1k}-1}{(X_{1k}-1)(N_{1k}-X_{1k})} + \dfrac{N_{0k}}{X_{0k}(N_{0k}-X_{0k})} \right)^{-1} + \sum\limits_{i=1,i\neq k}^{K} \hat{w}_i}$

and $g^{-1}(\hat{\psi}_{g,k}^b)$, $g^{-1}(\hat{\psi}_{g,k}^c)$ and $g^{-1}(\hat{\psi}_{g,k}^d)$, $k=1,...,K$ are defined analogously. Using similar arguments as in the proofs of the asymptotic properties of J_{MH}^{II}, consistency, asymptotic normality and asymptotic efficiency have been shown for the jackknife estimator J_{id}^{II} (Pigeot, 1989) with

$$J_{id}^{II} := N\hat{\psi}_{id} - \frac{N-1}{N} \sum_{k=1}^{K} \left(X_{1k}\hat{\psi}_{id,k}^a + (N_{1k}-X_{1k})\hat{\psi}_{id,k}^b + X_{0k}\hat{\psi}_{id,k}^c + (N_{0k}-X_{0k})\hat{\psi}_{id,k}^d \right).$$

That means desirable asymptotic properties are verified for the case, that the second jackknife approach is applied to the weighted arithmetic mean of the individually estimated odds ratios denoted by $\hat{\psi}_{id}$.

3. Finite-sample properties

To get an idea of the finite-sample properties of the above jackknife estimators and the classical estimators by Woolf (1955) and by Mantel and Haenszel (1959), a simulation study (Pigeot, 1990c) and a numerical comparison of the asymptotic relative efficiencies (Pigeot, 1990d) have been performed. Here a short description of the design of the studies and the most important results will be given.

Design of the Monte-Carlo study

In the simulation study, for each constellation of the involved parameters 1000 simulation runs were carried out. The number of contingency tables was chosen as 2, 5 and 10. Each of them was constructed with varying sample sizes N_{0k}, N_{1k}, $k=1,...,K$. The number of controls N_{0k} was always larger than the number of cases N_{1k} to reflect practical consideration. Besides the balanced case of a 1:3 ratio of cases and controls unbalanced samples were investigated. In addition to small sample sizes the tenfold of those were considered and referred to as moderate sample sizes. Five different values of the odds ratio

(1; 1.7; 3.5; 5; 10) were simulated connected with five different sets of p_{0k}, k=1,...,K; corresponding to 'narrow' differences of p_{0k} located close to zero, corresponding to 'wide' differences of the values of p_{0k} centered around 0.5, and corresponding to 'narrow' differences of p_{0k} centered around 0.5. The cases where the chosen values of p_{0k} were close to zero or one were only considered in connection with moderate sample sizes.

A similar design was used for calculating some asymptotic relative efficiencies.

Results of the simulation study

The Mantel-Haenszel estimator, the Woolf estimator, the transformed jackknife estimator of type I based on $\log(\hat{\psi}_{MH})$, denoted by $\hat{\psi}_{BL}$, the jackknife estimator of type II based on $\hat{\psi}_{MH}$, J_{MH}^{II}, and the transformed jackknife estimator of type I based on $\hat{\psi}_W$, referred to as $\exp(J_W^I)$, were then compared with respect to their sample bias, standard error, and mean squared error (MSE).

As already seen in other simulation studies, $\hat{\psi}_{MH}$ turns out to be preferable to $\hat{\psi}_W$ with respect to the bias (especially with small values of ψ) and $\hat{\psi}_W$ shows better results than $\hat{\psi}_{MH}$ concerning the MSE (especially with large values of ψ). This relation carries over when comparing the corresponding transformed jackknife estimators of type I. That means, $\hat{\psi}_{BL}$ is better than $\exp(\hat{\psi}_W)$ concerning the bias and with respect to the MSE this relationship is just vice versa. These four estimators are, however, dominated by J_{MH}^{II} in most of the situations considered. This estimator shows a remarkably smaller bias and MSE, especially with small sample sizes. While the transformed jackknife estimators of type I get better with an increasing number of tables, which is as expected, since they have been originally constructed for this case, the dominance of J_{MH}^{II} disappears with a large number of tables. In addition, J_{MH}^{II} shows the tendency to underestimate the true odds ratio, especially when a large number of tables contains only a few observations in the cells denoted by b or c. The other estimators tend to overestimate ψ.

Some general results may be stated, too. The standard error and the MSE of the above estimators increase with an increasing value of ψ and decrease, as it is expected, with increasing sample sizes or an increasing number of 2x2 tables. The absolute bias shows the same trends. No essential differences concerning the MSE and standard error of the estimators can be stated for the chosen sampling plans. In addition, the relationship among the estimators is not strongly influenced by the choice of p_{0k}. But it can be seen e.g., that J_{MH}^{II} shows very good results in connection with the two different choices of p_{0k} located close to zero. The Woolf estimator behaves very well concerning its MSE for a wide range of p_{0k}, especially for p_{0k} located close to zero or one. Here, $\exp(J_W^I)$ shows a deterioration. Further details concerning the simulation results are given in Pigeot (1990c).

Numerical comparison of the asymptotic relative efficiencies

The asymptotic relative efficiencies (ARE) were calculated for $\hat{\psi}_{MH}$, $\hat{\psi}_{BL}$ and $g(J_g^I)$, where the ARE of any estimator $\hat{\psi}$ of ψ is defined as $ARE(\hat{\psi}) := Var_A(\hat{\psi}_{ML})/Var_A(\hat{\psi})$ with $Var_A(\hat{\psi}_{ML})$ as asymptotic variance of the asymptotically efficient ML estimator derived by Gart (1962). The involved parameters were chosen as above. The results of this numerical comparison, which are described in detail in Pigeot (1990d), show that the AREs of all considered estimators are very high over a wide range of situations likely to occur in practice. The results are influenced by the values of the true odds ratio, the number of tables and the chosen values of p_{0k}, k=1,...,K, where only minor differences can be noticed concerning the chosen sampling plan. It can be stated that the ARE of $\hat{\psi}_{MH}$ as well as that of $g(J_g^I)$ is higher than that of $\hat{\psi}_{BL}$ in most cases. Moreover, the ARE of $g(J_g^I)$ is often higher than that of $\hat{\psi}_{MH}$ for large values of ψ and for a large number of tables.

The findings obtained for $\hat{\psi}_{MH}$ are similar to those described in Donner and Hauck (1986) and can be transferred to J_{MH}^{II}, because both estimators have the same asymptotic variance. Since the Mantel-Haenszel estimator is asymptotically efficient for $\psi=1$ (Tarone et al, 1983), it is not surprising, that $ARE(\hat{\psi}_{MH})$ is very high for small values of ψ. But this estimator shows also good results in other situations. In general, its ARE never drops below 0.86.

As already observed for the bias and MSE in the simulation study, the transformed jackknife estimators of type I $g(J_g^I)$ and $\hat{\psi}_{BL}$ behave also better concerning their ARE's for a large number of contingency tables. Thus, in the case of two 2x2 tables the ARE of $\hat{\psi}_{BL}$ is smaller than 0.9 in 28% of all situations and that of $g(J_g^I)$ in 46%. These estimators get especially poor with p_{01} chosen as 0.05 and p_{02} as 0.15 as well as with $\psi=10$, $p_{01}=0.05$ and $p_{02}=0.95$. With K=10, however, the ARE of $\hat{\psi}_{BL}$ is higher than 0.97 in 88% of all situations and that of $g(J_g^I)$ is always higher than 0.994.

4. Discussion

In Section 2, the asymptotic properties of some jackknife estimators have been discussed in the case when the sample sizes N_{0k} and N_{1k} within each table tend to infinity but the number K of 2x2 tables remains fixed. Some supplementary aspects should be mentioned according the proofs:

1. The asymptotic properties of $\hat{\psi}_{BL}$ can be shown also for the more general case that the first jackknife approach is applied to any function g^{-1} of $\hat{\psi}_{MH}$ with $g \in G$, where G is defined as in (1.4). Under the assumptions given above it can be verified that the transformed jackknife estimators $g(J_{g^{-1}(MH)}^I)$ with ‾

$$g(J^I_{g^{-1}(MH)}) := g\left(K \ g^{-1}(\hat{\psi}_{MH}) - \frac{K-1}{K} \sum_{k=1}^{K} g^{-1}(\hat{\psi}^I_{MH,k}) \right), \ g \in G,$$

are consistent and asymptotically normal with asymptotic expectation ψ and asymptotic variance $Var_A(\hat{\psi}_{BL})$.

2. In contrast to the asymptotic properties of $g(J^I_{g^{-1}(MH)})$ which hold for general choices of g as discussed in point 1, the proof of the consistency and asymptotic normality of the transformed jackknife estimators of type I based on $\hat{\psi}_g$ is still an open problem for other choices of g than the identity.

3. All asymptotics have only been investigated for increasing sample sizes but a fixed number of tables. The asymptotic behaviour of the above jackknife estimators in the alternative case of an increasing number of strata constitutes an interesting and also still open problem.

In Section 3, the estimators have been compared concerning their finite-sample properties as point estimators. Complementary to this study, they could be investigated with respect to their coverage probabilities when being used for the construction of confidence intervals.

References

ANSCOMBE, F.J. (1956). On estimating binomial response relations. *Biometrika* **43**, 461-4.

ARVESEN, J.N. (1969). Jackknifing U-statistics. *Ann. Math. Statist.* **40**, 2076-100.

BRESLOW, N.E. (1981). Odds ratio estimators when the data are sparse. *Biometrika* **68**, 73-84.

BRESLOW, N.E.; LIANG, K.Y. (1982). The variance of the Mantel-Haenszel estimator. *Biometrics* **38**, 943-52.

DAVIS, L.J. (1985). Weighted averages of the observed odds ratios when the number of tables is large. *Biometrika* **72**, 203-5.

DONNER, A.; HAUCK, W.W. (1986). The large-sample relative efficiency of the Mantel-Haenszel estimator in the fixed-strata case. *Biometrics* **42**, 537-45.

FLEISS, J.L.; DAVIES, M. (1982). Jackknifing functions of multinomial frequencies, with an application to a measure of concordance. *Am. J. Epidemiol.* **115**, 841-5.

GÄNSSLER, P.; STUTE, W. (1977). *Wahrscheinlichkeitstheorie*. Springer-Verlag, Berlin.

GART, J.J. (1962). On the combination of relative risks. *Biometrics* **18**, 601-10.

GART, J.J.; ZWEIFEL, J.R. (1967). On the bias of various estimators of the logit and its variance with application to quantal bioassay. *Biometrika* **54**, 181-7.

GASTWIRTH, J.L.; GREENHOUSE, S.W. (1987). Estimating a common relative risk: application in equal employment. *J. Am. Stat. Assoc.* **82**, 38-45.

GUILBAUD, O. (1983). On the large-sample distribution of the Mantel-Haenszel odds ratio estimator. *Biometrics* **39**, 523-5.

HALDANE, J.B.S. (1955). The estimation and significance of the logarithm of a ratio of frequencies. *Ann. Hum. Genet.* **20**, 309-11.

HAUCK, W.W. (1979). The large sample variance of the Mantel-Haenszel estimator of a common odds ratio. *Biometrics* **35**, 817-9.

HAUCK, W.W. (1989). Odds ratio inference from stratified samples. *Commun. Stat., Theory Methods* **18**, 767-800.

HAUCK; W.W.; ANDERSON, S.; LEAHY, F.J. III (1982). Finite-sample properties of some old and some new estimators of a common odds ratio from multiple 2x2 tables. *J. Am. Stat. Assoc.* **77**, 145-52.

HITCHCOCK, S.E. (1962). A note on the estimation of the parameters of the logistic function, using the minimum logit χ^2 method. *Biometrika* **49**, 250-2.

JEWELL, N.P. (1984). Small-sample bias of point estimators of the odds ratio from matched sets. *Biometrics* **40**, 421-35.

MANTEL, N.; HAENSZEL, W. (1959). Statistical aspects of the analysis of data from retrospective studies of disease. *J. Nat. Cancer Inst.* **22**, 719-48.

McKINLAY, S.M. (1975). The effect of bias on estimators of relative risk for pair-matched and stratified samples. *J. Am. Stat. Assoc.* **70**, 859-64.

McKINLAY, S.M. (1978). The effect of nonzero second-order interaction on combined estimators of the odds ratio. *Biometrika* **65**, 191-202.

NURMINEN, M. (1981). Asymptotic efficiency of general noniterative estimators of common relative risk. *Biometrika* **68**, 525-30.

O'GORMAN, T.W.; WOOLSON, R.F.; JONES, M.P.; LEMKE, J.H. (1988). A Monte Carlo study of three odds ratio estimators and four tests of association in several 2x2 tables when the data are sparse. *Commun. Stat., Simulation Comput.* **17**, 813-35.

PARR, W.C.; TOLLEY, H.D. (1982). Jackknifing in categorical data analysis. *Aust. J. Stat.* **24**, 67-79.

PIGEOT, I. (1989). Asymptotic properties of several jackknife estimators of a common odds ratio. *Technical Report 89/10*, Fachbereich Statistik der Universität Dortmund (submitted for publication).

PIGEOT, I. (1990a). A class of asymptotically efficient noniterative estimators of a common odds ratio. *Biometrika* **77**, 420-3.

PIGEOT, I. (1990b). A jackknife estimator of a combined odds ratio. To appear in *Biometrics*.

PIGEOT, I. (1990c). A simulation study of estimators of a common odds ratio in several 2x2 tables. To appear in *J. Statist. Comp. Simulation*.

PIGEOT, I. (1990d). Asymptotic relative efficiency of some jackknife estimators of a common odds ratio. To appear in *Biom. J.*.

SERFLING, R.J. (1980). *Approximation Theorems of Mathematical Statistics*. Wiley, New York.

TARONE, R.E.; GART, J.J.; HAUCK, W.W. (1983). On the asymptotic inefficiency of certain noniterative estimators of a common relative risk or odds ratio. *Biometrika* **70**, 519-22.

WOOLF, B. (1955). On estimating the relation between blood group and disease. *Ann. Hum. Genet.* **19**, 251-3.

AN APPLICATION OF THE BOOTSTRAP IN CLINICAL CHEMISTRY

Helmut Schäfer

Institut für Medizinische Biometrie und Informatik, Universität Heidelberg

1.   INTRODUCTION

We apply the bootstrap to a one-dimensional nonparametric discrimination problem in clinical chemistry: the construction of a cut-off point (discrimination limit) for a quantitative diagnostic test on the basis of a sample of "diseased" and "non-diseased" individuals and the subsequent evaluation of the resulting decision rule in the same sample. When the cut-off point is selected to maximize some performance criterion, the sample estimate of maximal performance is known to systematically overestimate the unknown true performance of the test at the selected cut-off point. Bootstrap methods are proposed in the classical paper by EFRON [2] to reduce bias and to obtain confidence intervals. We apply these methods and investigate their performance by a simulation study.

2.   THE APPLICATION PROBLEM

In patients with small cell lung cancer, the diagnosis of metastases governs the subsequent therapy. While patients without metastases (limited disease) have a small but existing chance for complete cure and therefore are treated with fairly aggressive chemotherapy plus radiotherapy, prognosis is very poor in the presence of metastases (extensive disease) and decision will be for less toxic therapy in favour of life quality. Liver and brain metastases can be diagnosed by CT scan or sonography, but for bone marrow metastases (BMM), the very unpleasant since invasive biopsy is performed. The question is: Can biopsy be replaced by a laboratory test? A good candidate is the tumour marker lactate dehydrogenase (LDH), which is produced by BMM (DOLL [1]). The upper part of Fig. 1 shows density estimates for log(LDH) in N_X=895 patients without (X) and in N_Y=95 patients with (Y) BMM, obtained from 3 randomized clinical trials ([4] [7]). We also consider the subsample (N_X=567, N_Y=27) of patients without any *other* metastases, which actually is the one relevant for the posed problem (in presence of other metastases, decision is clear without bone marrow biopsy!). The problem is to select a suitable cut-off point c for LDH: future patients with LDH < c will be classified as "no BMM" and the corresponding therapy will be started immediately, in those with LDH > c biopsy will be performed first.

3.   UTILITY FUNCTION AND SELECTION OF A CUT-OFF POINT

We will use $U(c) = 100 \cdot (1-p_Y) \cdot spec(c) - 100 \cdot p_Y \cdot 40 \cdot (1-sens(c))$ as an optimization criterion. Here, p_Y =prevalence of BMM, $spec(c) = P(X<c)$ (specificity) and $sens(c) = P(Y>c)$ (sensitivity). Note that $U(-\infty) = 0$ and U thus represents the utility gain compared to the optimal procedure without using the LDH-procedure, in which, according to the a priori probability, bone marrow biopsy is performed in every patient $(c=-\infty)$. U can be interpreted as the net number of biopsies saved per 100 patients, where 'net' means that 40 biopsies are subtracted for each overlooked BMM-patient. Plugging-in the estimators $\hat{p}_Y = N_Y / (N_Y + N_X)$, $\hat{spec}(c) = \hat{F}_X(c)$ and $\hat{sens}(c) = 1 - \hat{F}_Y(c)$ ($\hat{F}_X$ and $\hat{F}_Y$ = empirical distribution functions) gives a sample estimate $\hat{U}$, which is plotted in the lower part of Fig 1.

The cut-off point that maximizes $\hat{U}$ among all sample points is selected. We denote it by $\hat{c}_{opt}$. For the (567+27)-sample, we obtain $\hat{c}_{opt} = 5.36$ with utility $\hat{U}(\hat{c}_{opt}) = 28.5$. For the full sample, the corresponding values are 5.41 and 28.4.

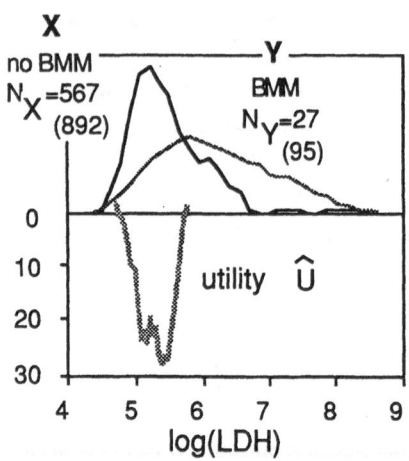

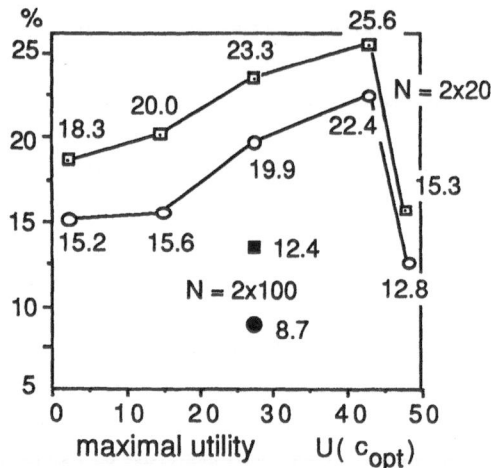

Fig. 1: Distributions of log (LDH) in patients with and without bone marrow metastases (BMM). Lower part: estimated utility (number of saved biopsies / 100 patients) of the test as a function of the cut-off value

Fig. 2: actual type I error of bootstrap confidence intervals.
$\square$ = two-sided, nominal $\alpha = 0.1$
O = one-sided, nominal $\alpha = 0.05$

4. THE BOOTSTRAP PROCEDURE

We denote by U, c_{opt}, F_X, F_Y the true utility function, the unknown true optimal cut-off point that maximizes U, and the true distribution functions, resp. The same letters with one hat, $\hat{U}$, $\hat{c}_{opt}$, $\hat{F}_X$, $\hat{F}_Y$, denote the estimators or the observed estimates in the present study (this will be clear from the context). The same letters with two hats, $\hat{\hat{U}}$, $\hat{\hat{c}}_{opt}$ denote the bootstrap random variables when X and Y are distributed according to $\hat{F}_X$ and $\hat{F}_Y$ and samples of the original sizes N_X, N_Y are drawn.

$\hat{U}(\hat{c}_{opt})$ can be viewed as an estimate for two different unknown quantitites, which should be carefully distinguished:
(A) the true utility $U(c_{opt})$ at the *unknown* true optimal cut-off point c_{opt}
(B) the true utility $U(\hat{c}_{opt})$ at the *known* cut-off point $\hat{c}_{opt}$ estimated from the given sample

To obtain corrected estimates and confidence limits for (A) or (B), resp., from the observed quantity $\hat{U}(\hat{c}_{opt})$, it would be necessary to know the distribution of the 'bias' $\hat{R}_A := \hat{U}(\hat{c}_{opt}) - U(c_{opt})$ or $\hat{R}_B := \hat{U}(\hat{c}_{opt}) - U(\hat{c}_{opt})$, resp. The corresponding bootstrap distributions are

(A) $\mathbf{D}(\hat{\hat{U}}(\hat{\hat{c}}_{opt}) - \hat{U}(\hat{c}_{opt}) \mid \hat{F}_X, \hat{F}_Y)$ (B) $\mathbf{D}(\hat{\hat{U}}(\hat{\hat{c}}_{opt}) - \hat{U}(\hat{\hat{c}}_{opt}) \mid \hat{F}_X, \hat{F}_Y)$

In type (A), the maximal utility in the bootstrap sample is compared with the maximal utility in the original sample. In type (B), the maximal utility in the bootstrap sample is compared with the utility from the original sample at the cut-off point optimized in the bootstrap sample. In contrary to type (A), here both terms vary from bootstrap replication to bootstrap replication, resulting in a larger variance. The means and the standard deviations of these distributions are the bootstrap estimators $\tilde{R}_A$, $\tilde{R}_B$ and $\widetilde{S.D.}$ $(\hat{R}_A)$, $\widetilde{S.D.}$ $(\hat{R}_B)$ for the means of $\hat{R}_A$ and $\hat{R}_B$ (i.e. the bias of $\hat{U}$ $(\hat{c}_{opt})$ with respect to U (c_{opt}) and U $(\hat{c}_{opt})$, resp.) and their standard deviations. Bootstrap estimators for U (c_{opt}) and U $(\hat{c}_{opt})$ themselves are obtained by $\widetilde{U(c_{opt})} = \hat{U}$ $(\hat{c}_{opt})$ $-\tilde{R}_A$ and $\widetilde{U(\hat{c}_{opt})} = \hat{U}$ $(\hat{c}_{opt})$ $-\tilde{R}_B$. Percentiles of distributions (A) and (B) yield bootstrap confidence intervals for U (c_{opt}) and U $(\hat{c}_{opt})$, resp.

By definition, type (B) estimates are appropriate for the practical application of the laboratory test at the given cut-off point $\hat{c}_{opt}$ in future patients. In a certain sense, type (B) is *constructive*, while type (A) only concerns the *existence* of a cut-off point with a certain utility.

The results are given in **table 1:**

type	sample	$\hat{U}$ $(\hat{c}_{opt})$	bias	S.D.	conf. interval	$\hat{c}_{opt}$	bias	S.D.	conf. interval
(A)	567+27	28.5	8.6	8.2	5.3 ... 31.8	5.36	-0.040	0.115	5.22 ... 5.55
(B)	567+27	28.5	10.9	9.5	0.2 ... 30.6	5.36	-0.040	0.115	5.22 ... 5.55
(B)	892+95	28.4	5.1	6.0	13.0 ... 32.6	5.41	0.021	0.107	5.22 ... 5.65

With 1000 bootstrap replications, the estimates are completely stable. For the smaller sample (567+27), the type (A) confidence interval clearly excludes a utility of 0 and thus states superiority of the LDH-test over the procedure of performing biopsy in every patient, while the type (B) confidence interval excludes 0 utility only with difficulty. In the larger sample, the type (B) bootstrap estimate (23.3) is almost 4 standard deviations away from 0. The one-sided bootstrap p-value for H_0: U $(\hat{c}_{opt}) \leq 0$ was p= 0.0034 (5000 replications). The estimated bias of the cut-off point $\hat{c}_{opt}$ itself is small. Here, the variance is more important.

5. A SIMULATION STUDY

To assess the validity of these results and the reliability of type B bootstrap for inferential purposes, a Monte Carlo study was performed with $X \sim \log-N$ $(0,1)$ and $Y \sim \log-N$ $(\mu,1)$ and U $(c) =50 \cdot$ $(\text{sens}$ $(c) +\text{spec}$ $(c) -1)$. Different degrees of overlapping of the X- and the Y-distribution were realized by taking $\mu=0.1, 0.75, 1.5, 3, 4$, which corresponds to a maximal utility U $(c_{opt}) =2$, 14.6, 27.3, 43.3, 47.7 (Note that $\sup U = 50$.). Equal sample sizes $N_X=N_Y=20$ were used to assess small sample behaviour (remember $N_Y=27$). To investigate the performance for larger samples, a simulation run with $N_X=N_Y=100$ (remember $N_Y=95$) was added for the case of $\mu=1.5$. This value was chosen for two reasons: The corresponding maximal utility of 27.3 is near to the one observed in our example, and the value of $\mu=1.5$ turned out to be the worst case in the first simulation run. For each of these parameter constellations, NSIM=1000 simulation trials were performed, with NBOOT=500 bootstrap replications per trial. A list of the random variables that were simulated can be obtained from the row headings of Tab. 2. For the notations, see section 4.

6.   RESULTS FOR POINT ESTIMATORS

The quantity one wants to estimate, $U(\hat{c}_{opt})$, is itself a random variable. Thus, the deviations of the estimators (resubstitution and bootstrap) from this variable were investigated and are tabulated in Tab. 2. In a similar Monte Carlo study for linear discrimination analysis in 2 dimensions with $N_X=N_Y=20$, EFRON [2] found a bias of 2.8 (S.D.10.3) for the resubstitution estimator and a bias of -0.1 for the boostrap estimator. In contrast, the bootstrap estimator is still biased in the case of one-dimensional nonparametric discrimination with the same sample size. The bias of both the resubstitution and the bootstrap estimator are decreasing functions of the degree of overlapping of X and Y (as measured by μ). The relative bias reduction achieved by the bootstrap (in Tab. 2: (2)-(3) divided by (2)) is a convex function of μ, varying from 55% to 67%. Standard deviations (Tab.2: in brackets) are slightly smaller for the resubstitution estimator, the mean square deviation from 0, $bias^2+S.D.^2$, however, is clearly smaller for the bootstrap estimator. Increasing sample size from 20 to 100 results in a 2/3 reduction of the bias of both estimators. Remarkably, the type B estimator, intended to estimate $U(\hat{c}_{opt})$ according to the bootstrap model, turns out to be an almost unbiased estimator for $U(c_{opt})$. In accordance with the findings of EFRON [2], $\widetilde{S.D.}(\hat{R}_B)$ comes out as a good (with regard to both its bias and variance) estimator for the S.D. of $\hat{R}_B$.

$N_X(=N_Y)$	20	20	20	20	20	100
μ	0.1	0.75	1.5	3.0	4.0	1.5
$U(c_{opt})$	2.0	14.6	27.3	43.3	47.7	27.3
(1) $U(\hat{c}_{opt})$	1.5	13.2	25.8	41.8	46.1	26.7
	(0.55)	(1.89)	(2.26)	(1.90)	(2.07)	(0.97)
(2) $\hat{U}(\hat{c}_{opt})-U(\hat{c}_{opt})$	8.7	7.3	6.1	3.8	3.0	2.4
	(5.46)	(6.15)	(5.63)	(3.60)	(2.59)	(2.35)
(3) $\widetilde{U}(\hat{c}_{opt})-U(\hat{c}_{opt})$	3.2	2.4	2.0	1.4	1.3	0.7
	(6.16)	(7.04)	(6.40)	(4.24)	(2.89)	(2.61)
(4) $\widetilde{S.D.}(\hat{U}(\hat{c}_{opt})-U(\hat{c}_{opt}))$	5.7	5.9	5.3	3.2	2.4	2.7
	(0.49)	(0.40)	(0.55)	(0.75)	(0.42)	(0.15)

Table 2:   Means and S.D (in brackets) of the following random variables.
        (1): unknown actual utility at the selected cut-off point
        (2): deviation of the resubstitution estimator from (1)
        (3): deviation of the bootstrap estimator from (1)
        (4): bootstrap estimator for the S.D. of (2).

7.   RESULTS FOR CONFIDENCE INTERVALS

Since in our application we are mainly interested in statistical inference, we also determined the actual confidence level of two-sided bootstrap confidence intervals with the nominal level of 90%, and of one-sided lower confidence intervals with a nominal level of 95%. Note that it is exactly the lower bound which is relevant for our application problem, in which one wants to infer that the true utility is > 0. Fig.2 shows the actual type I error risk as a function of the distribution parameter μ (converted into the true maximal utility $U(c_{opt})$).

First, the nice property of the curves: they have an upper bound. Thus, one could try to adjust the nominal α to tie down the complete curve beyond the desired level, which then would be held over the whole range of the unknown distribution parameter μ. We did not follow this idea in the present paper. For a similar method, see SCHÄFER [6]. The bad property is the fact that, for the sample size of 20, in the maximum, the actual type I error α is 2.5 times the nominal value. The α-inflation is reduced from the factor of 2.5 to a factor of less than 1.25 by increasing the sample size to 100. However, this joy is spoiled by the fact that, due to the bias, the remaining 12,4% are not equally distributed to both sides. Thus, for the one-sided lower confidence interval with nominal $\alpha=5\%$, the actual α is 8,7% even for the larger sample size of N=100, which is still unsatisfactory.

8.   DISCUSSION AND CONCLUSIONS

Besides the very inefficient procedure of sample splitting, bootstrap techniques seem to be the only existing approach for statistical inference on the utility $U(c_{opt})$ or (more relevant) $U(\hat{c}_{opt})$. Other approaches in the literature do not meet the problem. Gail and Green [3] determine the distribution of $\hat{U}(\hat{c}_{opt})$ under $F_X=F_Y$. The alternative $F_X<F_Y$, however, does not imply positive utility $U(c_{opt})>0$, except for the special choice $U(c)=spec(c)+sens(c)-1$, which will be inadequate in general. Likewise, the method of Miller and Siegmund [5] is nothing but an unconventional and not very powerful test of H_0: $F_X=F_Y$.

There are 2 types of bootstrap estimators and bootstrap confidence intervals, those referring to the unknown true optimal cut-off point c_{opt} and those referring to the special known estimate $\hat{c}_{opt}$. The second one should be used in practice. The bootstrap removes about 60% to 70% but not all of the bias of the resubstitution estimator. Especially for one-sided lower bootstrap confidence intervals, error risks are relevantly increased over the nominal value, even for sample sizes of 2 x 100. No rules for α-adjustment are available. Concerning the LDH-example, the one-sided bootstrap p-value of 0.0034 obtained for the full sample is probably small enough to exclude actual error risks of more than 0.05. Thus, if one is willing to use this sample (see the remark on appropriateness in section 2), this allows to establish a concrete cut-off point that warrants utility > 0 for the LDH-test.

REFERENCES
[1] Doll, D.C.:'Serum lactate dehydrogenase and bone marrow involvement in small cell carcinoma of the lung' *New England Journal of Medicine* **12**, p. 1262 (1985)
[2] Efron, B.:'Bootstrap methods: another look at the jackknife' *The Annals of Statistics* **7**, 1-26 (1979)
[3] Gail, MH, Green, SB: 'A generalization of the one-sided two-sample Kolmogorov-Smirnov statistic for evaluating diagnostic tests' *Biometrics* **32**, 561-570 (1976).
[4] Havemann, K., Wolf, M., Holle, R. et al.:'Alternating vs. sequential chemotherapy in small cell lung cancer. A randomized German multicenter trial' *Cancer* **59**, 1072-1082 (1987)
[5] Miller, R, Siegmund, D: 'Maximally selected chi-squared statistics'*Biometrics* **38**, 1011-1016 (1982).
[6] Schäfer, H.:'Constructing a cut-off point for a quantitative diagnostic test' *Statistics in Medicine* **8**, 1381-1391 (1989)
[7] Wolf, M., Havemann, K., Holle, R. et al.:'Cisplatin/Etoposide versus Ifosfamide/Etoposide combination chemotherapy in small cell lung cancer: a multicenter German randomized trial' *Journal of Clinical Oncology* **5**, 1880-1889 (1987)

Applications of Monte-Carlo-Techniques (9)

COMPUTER AIDED SIMULATION ANALYSIS (CASA)
IN ORDER TO TEST COST-EFFECTIVENESS OF DRUGS USING THE EXAMPLE
OF CEFTRIAXONE (ROCEPHIN[R])

Rito Bergemann, Arno Brandt, Thomas Nawrath,
Alexander Richter, Walter Siegrist, Fred Sorenson

Institute for Medical Informatics Basel, Bachtelenweg 3, CH-4125 Riehen

I. Introduction

It is well known that the methods currently being used in successful drug development are very expensive, involving a large number of carefully monitored clinical trials and provide the results only after a long period of time. Even when a standardized registration procedure is followed, however, it does not necessarily guarantee that the medication will become available on the market. This may be due in part to the pharmaceutical industry which, in addition to fulfilling its usual requirements in establishing sufficient proof of the efficacy and safety of a new drug to the regulatory authorities as well as to the health insurances, also often has to substantiate the drugs economic efficiency.

The CASA (Computer Aided Simulation Analysis) is a new method developed to compare the economic efficiency of different drug therapies. It is used to test clinical and/or socioeconomic hypotheses derived from clinical data on the drugs or therapies to be compared. This approach allows one to explore "in vitro" the effects of bringing together clinical, epidemiological and socioeconomic information. The clinical scenario is structured into a decision tree, whereby possible courses of events, actions and outcomes of treatments are depicted explicitly and transparently. The potential clinical pathology of patients is described by using transition probabilities. The pathway of a simulated patient through the structured scenario is selected by chance within the frame of given frequency distributions.

Ceftriaxone is a potent antimicrobial drug. Hitherto, its application in Switzerland was restricted to hospitals. The extension of the application area to ambulatory care was subject to authorization by the federal administration. The aim of this study was to analyze the economic effects of a prospective admission of ceftriaxone to include a limited spectrum of infectious diseases in ambulatory care. Specifically, admission of ceftriaxone for out-patient treatment of two particular infectious diseases (bronchopneumonia and pyelonephritis) was thought to be worthwhile, because a significant reduction of hospitalizations could be expected.

A model for ambulatory and stationary treatment of bronchopneumonia and pyelonephritis was designed. This model then evaluated treatment with and without ceftriaxone in order to compare the economic consequences of both strategies. With the aid of the Monte Carlo method used by the computer software which we employed, it was possible to generate a realistic view of the outcomes of the treatments. The simulation technique not only allowed us to statistically compare the target variables, but also provided important information of the distribution of individual costs within the study populations.

II. Methods

Since the primary objective of the analysis consisted of a comparison between two alternative strategies, the model we designed was based on a decision tree involving two main branches as depicted using the Jackson diagram in Figure 1. The first main branch represents the present strategy (strategy A), that is ceftriaxone is not licensed for ambulatory treatment of bronchopneumonia or pyelonephritis and in the second branch (strategy B), ceftriaxone is admitted as a potential treatment for these indications.

The simulated patients which were run through the model suffered either from bronchopneumonia or from pyelonephritis. The relative probabilities associated with the two infections evaluated were calculated on the basis of morbidity data. The percentage of hospitalizations in strategy A were taken directly from the available data and it was hypothesized and incorporated into the model that the number of patients who required stationary treatment would be 1.5% lower with strategy B than with strategy A. With respect to the costs of ambulatory treatment in strategy B, the decision tree was further divided into two subbranches representing treatment with either ceftriaxone or a conventional drug. The percentage of therapy substitution used in calculating this part of the model was again hypothesized. In all cases where hypothesized data were fit into the model, special care was taken to ensure that only the most conservative assumptions with respect to ceftriaxone were used.

Following construction of the model, 100,000 simulated patients were run through each strategy of the decision tree. When the endpoint of the tree was reached, the total costs, i.e. the sum of all diagnostic and therapeutic procedures, resulting from the treatment of each patient was then determined. These costs consisted of components which were allowed to vary within defined boundaries and in accordance with given distributions. The distribution function of each variable was either derived from field surveys or the appropriate literature, whichever was most applicable. Tables were made which contained the probability mass function values from this data. From these tables, the cumulative mass functions were derived by means of smoothing, integration and normalization. The population was divided into three age groups, taking into account the fact that the duration of hospitalization is age dependent.

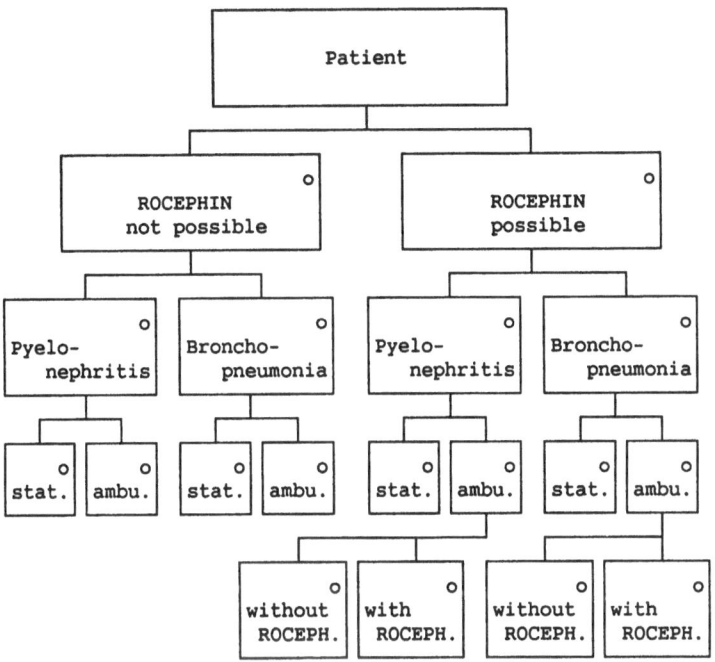

Figure 1: Design of the decision model

The patient number calculation uses the following method: The first number of patients was 1'000 and the second 10'000. In the case of a relative difference between the cost results of these simulations greater than 1% of the results with the lower number of simulation runs we do another simulation with 10'000 and 10'0000 patients. Between the simulations with 100'000 and 1'000'000 patients the simulation shows a difference for the costs lower than 1%.

For the description of the results we used methods of descriptive statistics. All analysis were performed with the statistical software package SAS Version 6.03 for PC.

The generation of the decision tree, the calculation of the distribution functions and the execution of the simulations on personal computers were supported by the software SMLTREE [R] (programmed by Jim Hollenberg). This software runs on every PC

with the operating system DOS 3.0 or higher and needs no specific graphic adapter
cards.

III. Results

A total of 100,000 simulated patients for each strategy with either bronchopneumonia
or pyelonephritis were run through the decision tree model described in the
methodology section. Table 1 summarizes the results of the strategy which included
the admission of ceftriaxone in out-patient treatment and Table 2 provides the
results of the alternative strategy where ceftriaxone was not admitted. Both tables
show the number of patients evaluated and the average and total costs of all
possible outcomes.

Table 1: Strategy B: Admission of Ceftriaxone

endpoint	number	costs(avg)	total costs
bronchopneumonia:			
ambulant:			
ceftriaxone	3,045	1,155	3,515,957
other	63,643	769	48,953,254
stationary:	9,220	7,954	73,337,384
pyelonephritis:			
ambulant:			
ceftriaxone	550	1,107	608,596
other	21,672	762	16,509,882
stationary:	1,870	6,795	12,707,269
	100,000	1,556	155,632,342

Table 2: Strategy A: Ceftriaxone not admitted

endpoint	number	costs(avg)	total costs
bronchopneumonia:			
ambulant:	65,949	769	50,700,760
stationary:	10,354	7,946	82,268,336
pyelonephritis:			
ambulant:	21,725	762	16,549,338
stationary:	1,972	6,739	13,289,876
total	100,000	1,628	162,808,310

When viewing the simulated patients as a whole, ambulatory treatment with
ceftriaxone resulted in additional costs when used versus bronchopneumonia
(SFr. 1,768,451) as well as pyelonephritis (SFr. 569,140). On the other hand,

reduction of the hospitalization rate by using this drug was projected to save SFr. 8,930,952 (bronchopneumonia) and SFr. 582,607 (pyelonephritis). These results suggest that a substantial benefit, financially speaking, would be achieved if ceftriaxone was admitted as treatment in the 100,000 simulated patients for these two infectious diseases.

IV. Discussion

This study has demonstrated an example where a decision tree model, in combination with the Monte Carlo simulation technique, could be used as a valuable tool to support decisions in the field of general health care. Although the CASA necessarily simplified the clinical situation, the advantages of such a method are readily apparent. It was able to produce an artificial clinical/economical trial with its resulting data subsequently analyzed using standard biometric methods. The study duration was short and new simulations based on the latest available information or controlling for certain factors can be easily implemented. In addition, the model allowed us to make predictions, not only on the expected values, but also on the distribution of the target variables. It was also possible to determine where one particular strategy is most optimal.

Since there may exist some uncertainty about the transition probabilities in a decision tree, the model we used would also support second order Monte Carlo analysis, also referred to as probabilistic sensitivity analysis. In this case, the probability itself could be randomly selected within a given interval and according to a given distribution function. Our model could also include the possibility to simulate time-dependent processes whereby the tree nodes could be arranged in a recursive way and the parameters set to functions of time. This would make it possible to integrate Markow chains into the decision tree.

V. References

Arzneimittel-Kompendium der Schweiz 1989. Documed AG, Basel, 1989

Economics of ceftriaxone use. Editiones <Roche>, Basle, 1987

Fernex, M., Havas, L., Ryff, J.C., Clarke, M.R.: Ergebnisse der klinischen Prüfung von Ceftriaxon. Hahnenklee-Symposium, Basel 1981: Editiones <Roche>, 1982, pp. 389-415

International Classification of Diseases, 9th Revision, U.S. Departement of Health and Human Services, DHHS Publication No. (PHS) 80-1260

Prognos-Beiträge zur Unternehmungsführung. Simulation als Instrument der Planung in Wirtschaft und Verwaltung. Verlag moderne Industrie, Zürich, 1973

Schweizer Spital, 52. Jahrgang Heft 11, 1988

Schweizerischer Diagnose Index. Institut für medizinische Information und Statistik (IMS), Zug, 1988

Spezialitäten-Liste. Bundesamt für Sozialversicherung (BSV), Bern, März 1989

VESKA Medizinische Statistik. Gesamtstatistik 1987 Diagnosen und Operationen Vereinigung Schweizerischer Krankenhäuser (VESKA), Aarau, 1988; pp 55-61

Doubilet, P., Begg, C.B., Weinstein, M.C., Braun, P., McNeil, B.J.: Probabilistic Sensitivity Analysis Using Monte Carlo Simulation. A Practical Approach. Medical Decision Making 1985, Vol.5, No.2, pp 157-177

SMLTREE : computer program for decision analysis and Monte Carlo Simulation Jim Hollenberg, 445 E 68th St, BOX 20, New York, NY 10021, USA

Some Modelling and Simulation Issues Related to Bootstrapping

Mark E. Johnson
Professor and Chair
Department of Statistics
University of Central Florida
Orlando, Fl, USA 32816-0370

Abstract

The paper presented at the conference was based primarily upon "Estimating Model Discrepancy," Technometrics, 1990, co-authored with Shane Pederson. Since this article is easily available at University libraries, the present paper will focus on the chronology of work leading up to the Technometrics article. Further, some basic simulation and modelling issues related to bootstrapping are presented.

1. Introduction.

Let us be brutally honest. Most Ph.D. dissertations are not intended to nor are they expected to provide original results that will have future application to a real problem or data set. In fact, many of us can pursue research directions for years without need of a reality-check, since the technical details can be so utterly entertaining.

I started my research career in the area of multivariate distributions and their simulation. At the time, researchers were devoting considerable efforts at refining existing univariate algorithms to become exceedingly fast. My concern was not how fast are my generators, but what I could and should be generating in the multivariate setting.

Certain "classical" distributions naturally lend themselves to variate generation. The Johnson translation system is particularly direct to handle. The spherically symmetric class, which includes the multivariate normal distribution, can be very easy. For details, see Johnson (1987). For the better part of ten years I found these distributions to be of sufficient interest to maintain my activity. With the culmination of my favorite results in book form, I was ready to "retire" from this area of research and look for another domain. All would have been fine, if only ...

2.   The Problem.

In September of 1985, John McKee, a physicist/engineer wandered into my office with a question related to a particle scattering problem (neutral particle beam application). Initially, the problem as posed suggested that particles were being deflected from their true course by an angle having a Gaussian distribution (physicists resist the term normal in deference to Gauss, I guess). After several weeks of conversations and joint cross-training in each other's disciplines, we arrived at a model that captured the physical situation. The particles were distributed on the target plane according to some spherical distribution, possibly the Gaussian, but more likely some form of the t-distribution. Physicists knew two of the special cases as Lorentz scattering and Coulomb shielded scattering distributions.

Once some potential models were available, the next issue became the relationship of these models to actual data, if any. We tracked down some ancient experiments and roughly forty data sets to play with. We could settle the issue of which model is right!

First, however, we needed to develop the estimation scheme for the data sets. As happens so often, it was not a trivial matter to estimate the parameters in these distributions. The data had been grouped already in bins (over which we had no control) and the data was truncated with unknown counts of particles in the tails. Using a grouped maximum likelihood approach, Beckman and Johnson (1986) arrived at a relatively automatic scheme to fit the forty data sets. This paper holds my personal best record for time from start of problem until appearance in print. Everyone would have lived happily ever after, if only ...

3. More Problems

The results of the forty fits were disturbing in two ways. First, no "one" model worked over the forty different data sets. In one experiment, the fitted distribution had tails heavier than a Cauchy; in others, the tails were lighter than a t-distribution with 4 degrees of freedom. Second, the quality of the fits, as measured by a chi-square goodness-of-fit statistic had the obnoxious habit of "rejecting" the fits in cases with 10,000 particles or more, although visually, the fits were practically acceptable. For data sets with only hundreds of particles, the fits looked unacceptable, but passed the test. (I should add here that the physicists were very comfortable with chi-square--they had in fact been using an estimation package from Switzerland based on minimum chi-square). Of course, the second dilemma is well-known.

Berkson (1938) pointed out the problem more than fifty years ago. It remains painfully embarrassing, however, when objective measures fly in the face of common sense.

About this time, Pederson entered the scene equipped with extensive experience in categorical data analysis and an awareness of the wonderful paper of Cressie and Read (1984). To make a long story short and avoid repeating the paper by Pederson and Johnson (1990), we were able to modify X^2/N, which is the usual adjustment to chi-square to handle Berkson's problem, to behave correctly in small-samples with sparse data situations. At long last, we were able to avoid the statistically embarrassing situation -- the plots and our modified X^2/N statistic were consistent.

4. Conference on Bootstrapping and Related Techniques

Being one of the few presenters on the topic "Related Techniques," I felt compelled at the conference to provide at least some opinions on bootstrapping, as it relates to some practical issues of assessment in small samples. The basic theme of these remarks is that as the mainline bootstrappers descend from the asymptotic stratosphere, I hope that they will avoid certain pitfalls.

An old adage goes that a lawyer who represents himself has a fool for a client. A similar indictment could be made for various statisticians who have designed their own Monte Carlo studies in support of theoretical investigations. To find examples of confounding that undercut conclusions in scientific investigations, the statistics journals on my shelves are a fertile place to look.

Aside from the previously mentioned references, Ripley (1987) and Kleijnen (1987) warrant consultation.

An aspect of the particle scattering problem not mentioned above but relevant at this point concerns the particle counts. Had the ancient experiments in fact collected an arbitrarily large (even infinite) number of particle counts, there was still no information there about the validity of the spherical symmetry. Translation: an infinite amount of information about the wrong issue can tell us nothing about the problem at hand. I hope bootstrappers will take into account the pertinent issues at hand and collect data accordingly. Frankly, I am a little nervous that in the zeal for rapid publication, corners could be cut on the empirical investigations.

5. Main Conclusion.

The experiences chronicled here indicate that in one individual's situation, the appearance of a real problem with real data revived and enhanced his research in the area. Real **new** data sets provide the possible inspiration for new methodology. Several classical data sets (Brownlee's stack loss data, ball-bearing failures, Canadian lynx furs trapped from 1821-1934) survive the test of time to the point where one wonders if **new** procedures have been developed which only work well on one of these data sets. Genug ist schon genug (Enough is enough already.)

References.

Beckman, R.J. and Johnson, M.E. (1987). "Fitting the Student-t Distribution to Grouped Data, With Applications to a Particle Scattering Experiment," Technometrics, 29, 17-22.

Berkson, J. (1938). "Some Difficulties of Interpretation Encountered in the Application of the Chi-Square Test," Journal of the American Statistical Association, 33, 426-442.

Cressie, N. and Read, T.R.C. (1984). "Multinomial Goodness-of-Fit Tests," Journal of the Royal Statistical Society, Series B, 46, 440-464.

Johnson, M.E. (1987). Multivariate Statistical Simulation. New York: Wiley.

Kleijnen, J.P.C. (1987). Statistical Tools for Simulation Practitioners. New York: Dekker.

Pederson, S.P. and Johnson, M.E. (1990). "Estimating Model Discrepancy," Technometrics, 32, 305-314.

Ripley, B.D. (1987). Stochastic Simulation. New York: Wiley.

SUPERCOMPUTERS FOR MONTE CARLO SIMULATION:

CROSS-VALIDATION VERSUS RAO'S TEST

IN MULTIVARIATE REGRESSION

Jack P.C. Kleijnen
Katholieke Universiteit Brabant (Tilburg University)
Postbox 90153, 5000 LE Tilburg, Netherlands
Bitnet: kleijnen@kub.nl   Fax: +3113-663072

Abstract

Part I covers vector computers, in the context of Monte Carlo experiments with regression models. These computers should exploit a specific dimension of the Monte Carlo experiment, namely its replicates. The resulting code computes Ordinary Least Squares (OLS) estimates on a CYBER 205 in 2% of the time needed on a Vax 8700. For Generalized Least Squares, however, the code runs slower on the CYBER 205 if the regression model is small; for large models the CYBER runs much faster. Part II covers regression models with correlated errors. To test the validity of the specified regression model, Rao (1959) generalized the F statistic for lack of fit, whereas Kleijnen (1983) proposed cross-validation using Student's t statistic and Bonferroni's inequality. A large Monte Carlo experiment compares these two methods, for normal and non-normal errors. Under normality, cross-validation is conservative, whereas Rao's test realizes its nominal type I error and has high power. Several confidence interval procedures for regression parameters are also compared. Under lognormality, only cross-validation with OLS works.

Part I: Vector computers

Part I is based on Kleijnen and Annink (1990). It illustrates three important points about "supercomputers":

(i)   Efficient supercomputing requires that algorithms be adjusted to take advantage of the specific architecture of the computing hardware.

(ii)   For certain problems, expensive supercomputers may be slower than general purpose machines are.

(iii) The increased speed of the supercomputers may not outweigh the burden of constructing the specialized code: the researcher's time is valuable too.

Vector computers are to be distinguished from traditional scalar computers and truly parallel computers. Traditional computers such as the IBM 370 and the VAX series, execute one instruction after the other; so they operate sequentially. Truly parallel computers such as the HYPERCUBE, have many Central Processing Units (CPU's) that can operate independently of each other; this is called coarse grain parallelism. Vector computers such as the CRAY 1 and the CYBER 205, have a "vector processing" capability: fine grain parallelism. Consider, for example, the computation of the inner product of two vectors: $v_1' v_2 = \Sigma_{j=1}^{n} v_{1j} v_{2j}$. This computation requires n identical scalar operations $v_{1j} v_{2j}$. The vector processor starts computing $v_{1j} v_{2j}$ while the computation of the predecessors $v_{1(j-1)} v_{2(j-1)}$, $v_{1(j-2)} v_{2(j-2)}$, $\cdots$ is still in process! So a vector computer works as an assembly line. A technical condition is that the scalar operations do not depend on each other; in the example the computation of the scalar product $v_{1j} v_{2j}$ does not need the other scalar products. Such an architecture is called a pipeline. The pipeline or assembly line requires a fixed set up cost; consequently a vector computer works efficiently only if a "large" number of identical scalar operations can be executed independently of each other. In Part I we show how to formulate the Monte Carlo model such that a vector computer can be applied efficiently. Technical details on the new generation of "supercomputers" are given by Levine (1982); more references are provided in Kleijnen and Annink (1989).

1. Regression Models and Simulation

Consider the well-known linear regression model

$$E(\mathbf{y}) = X \, \beta \tag{1.1}$$

with $y = (y_1, \ldots, y_i, \ldots, y_n)'$, $\beta = (\beta_1, \ldots, \beta_j, \ldots, \beta_Q)'$ and $X = (x_{ij})$ where $i = 1, \ldots, n$ and $j = 1, \ldots Q$. We assume additive errors $e = (e_1, \ldots, e_i, \ldots, e_n)'$:

$$y = X\beta + e. \tag{1.2}$$

We further assume that e is n-variate normally (N_n) distributed:

$$e \sim N_n [0_n, \text{cov}(e)], \tag{1.3}$$

where 0_n denotes a column of n zeros; $\text{cov}(e)$ denotes the variance-covariance matrix of e; $\text{cov}(e)$ equals $\text{cov}(y)$ because of (1.2); $\text{cov}(y)$ is assumed to be nonsingular.

This regression model can be applied to analyze the results of a simulation experiment that uses the *same* pseudorandom number sequence; see Kleijnen (1987) and Kleijnen (1988). In a well designed simulation experiment it is easy to *replicate* each factor combination; that is, row i of X is observed $m \geq 2$ times. Unbiased estimators of $\sigma_{ih} = \text{cov}(y_i, y_h) = \text{cov}(y_{ir}, y_{hr})$ where y_{ir} is the rth replication of the ith factor combination, are:

$$\widehat{\text{cov}}(y) = (Y\,Y' - \bar{y}\,\bar{y}'m)/(m-1), \tag{1.4}$$

with $\widehat{\text{cov}}(y) = (\hat{\sigma}_{ih})$, $Y = (y_{ir})$, $\bar{y} = (\bar{y}_i)$ with the averages $\bar{y}_i = \Sigma_{r=1}^m y_{ir}/m$; by definition we have $\sigma_{ii} = \sigma_i^2$. It is simple to prove that $\widehat{\text{cov}}(y)$ is singular for $m \leq n$.

We consider two different point estimators for $\underline{\beta}$. The classic estimator uses *Ordinary Least Squares* or OLS:

$$\hat{\beta} = (X'X)^{-1}X'\bar{y} , \tag{1.5}$$

which assumes $n \geq Q$ and rank $(X) = Q$. If $\text{cov}(y)$ were known, then a better estimator would use Generalized Least Squares (GLS). Since $\text{cov}(y)$ is unknown in practice, we replace it by the estimator $\widehat{\text{cov}}(y)$ of (1.4), and use *Estimated Generalized Least Squares* or EGLS:

$$\tilde{\beta} = (X'\widehat{\text{cov}}(y)^{-1}X)^{-1}X\,\widehat{\text{cov}}(y)^{-1}\bar{y} , \tag{1.6}$$

which assumes that $\widehat{\text{cov}}(y)$ is non-singular.

2. Vectorizing the Monte Carlo program

We wish to construct a Monte Carlo program to compare the OLS and EGLS estimators of (1.5) and (1.6). We might use the vector mode to compute an individual element y_{ir} of Y; see (1.2). But X is n x Q; typically n and Q range between n = 8 and Q = 4 and n = 32 and Q = 22. Vector computers are inefficient if the number of parallel operations is "small", say, smaller than 50; see Levine (1982) and SARA (1984). So it is inefficient to vectorize the computation of an individual y_{ir}.

Next we consider the vectorization of either the rows or the columns of Y. Since there are only n rows (factor combinations), vectorization is again inefficient. Because the m columns of Y are statistically independent (see § 1), vectorization is possible. In practice, however, m will be small (the minimum is m = n + 1; otherwise, $\hat{cov}(y)$ is singular). So vectorizing the columns of Y is also inefficient.

The Monte Carlo experiment is replicated L = 100 times, as we shall see in Part 2. (These Monte Carlo replicates ℓ with ℓ = 1,...,L, must be distinguished from the simulation replicates r = 1,...,m.) The L Monte Carlo replicates are statistically independent; they can be vectorized as we show now. Note that the more replicates we wish to obtain, the more efficient the vector computer becomes. Vectorization may be compared to filling a three-dimensional box in parallel with errors $e_{ir\ell}$ with i = 1,...,n; r = 1,...,m; ℓ = 1,...,L.

Step 1: Sample pseudorandom numbers x in parallel

Kleijnen (1989) evaluates several procedures for the parallel generation of pseudorandom numbers x ~ U(0,1). Kleijnen and Annink (1989) recommend the following generator. Take a scalar multiplicative congruential generator with a multiplier that gives acceptable statistical behavior. To initialize the vector version of this generator, first generate - in scalar mode - a vector of J successive pseudorandom integers x = $(x_0, x_1, x_2, \ldots, x_{J-2}, x_{J-1})'$ with seed x_0 and $x_j = (a\, x_{j-1})$ mod m for j = 1,2,..., J-1. To obtain numbers between zero and one, divide by m. Once and for all, compute a scalar multiplier (a^J) mod m. Vector multiplication of the vector x with this scalar multiplier gives a new vector: $(x_J, x_{J+1}, \ldots, x_{2J-2, 2J-1})'$. In this way the pseudorandom numbers are generated in parallel, in exactly the same order as they would have been produced in scalar mode. At the end of the Monte Carlo experiment the vector of the last J numbers should be stored, so that the experiment can be continued later on.

We mentioned that vector computers become more efficient as the number of parallel operations increases. For the CYBER 205, however, there is a technical upper limit: J = 2^{16} - 1 = 65 535 (since this computer uses 16 bits for addressing; see SARA, 1984, p. 26).

There is a computational problem: overflow occurs when computing (a^J) mod m. This problem is solved through the techniques of "controlled integer overflow" and the CYBER 205's "two's complement" representation of negative integers; see Kleijnen and Annink (1989).

Step 2: Sample independent standard normal variates z in parallel

There are many techniques for generating normal variates. We select the following procedure that fits a vector computer:

$$z_1 = (-2 \ln x_1)^{\frac{1}{2}} \cos 2\pi x_2 \qquad\qquad (2.1.a)$$

$$z_2 = (-2 \ln x_1)^{\frac{1}{2}} \sin 2\pi x_2, \qquad\qquad (2.1.b)$$

where the mutually independent pair x_1 and x_2 with $x \sim U(0,1)$ yields the mutually independent pair z_1 and z_2 with $z \sim N(0,1)$. To compute the functions $\ln$, cos, and sin for a *vector* of numbers, we use FORTRAN 200's vector functions VLN, VCOS, and VSIN. Given a vector of L independent pseudorandom numbers x, we use the first half to compute L/2 independent parallel realizations of $\ln x_1$, and the second half to compute cos $(2\pi x_2)$ and sin $(2\pi x_2)$: Figure 1 gives a pseudo-FORTRAN program where π is computed through the arccosine function; see SARA (1984, p. 13). To convert this pseudo-FORTRAN into a FORTRAN 200 program, we can replace DO loops by the special syntax of FORTRAN 200; the supercomputer can also automatically translate the FORTRAN program of Figure 1 provided we add CONTINUE statements; see SARA (1984, p. 17).

FIGURE 1

Parallel computation of L variates $z \sim N(0,1)$.

```
      L2 = L/2; PI = ACOS(-1.0); C = 2 * PI
      DO    20    LL = 1, L2
20          HELP1(LL) = SQRT(-2 * LOG(X(LL)))
      DO    30    LL = 1, L2
            HELP2(LL) = COS(X(LL + L2) * C)
            HELP3(LL) = SIN(X(LL + L2) * C)
            Z(LL) = HELP1(LL) * HELP2(LL)
30          Z(L2 + LL) = HELP1(LL) * HELP3(LL)
```

Above we saw that we wish to fill a three-dimensional "box" with $e_{ir\ell}$. So we store the *vectors* z (with L elements) of Figure 1 into a three-dimensional *array*

$Z(i,r,\ell)$.

Step 3: Sample n-variate normally distributed errors e

The columns of the matrix Y are n-variate normal. The sampling subroutine for multi-variate normal e with covariance matrix cov(e) is

$$e = C\ z, \tag{2.2}$$

with $z = (z_1, \ldots, z_i, \ldots, z_n)'$ and independent $z_i \sim N(0,1)$, and C a lower triangular matrix defined by

$$C\ C' = cov(e), \tag{2.3}$$

which is computed by Choleski's technique; see standard software libraries such as IMSL and NAG. Once C is computed, we generate e through the linear transformation (2.2) of z; that transformation is not vectorized because n is too small.

To obtain M observations and L Monte Carlo replicates of e, we use the FORTRAN program of Figure 2, where M denotes the maximum value of m in the experiment (here M = 33) and E(LL,R,I) is zero initially. C or C(I,J) does not vary over seeds (R) and Monte Carlo replicates (LL); it does vary over the Monte Carlo experiments de-fined by cov(y). To vectorize a program we should make the *inner* DO loop long (see the LL loop); moreover we should store the columns of the array columnwise; see SARA (1984, pp. 15, 20-21, 33).

FIGURE 2

Vectorized FORTRAN program for e.

```
      DO    20    I = 1,N
        DO    20    J = 1,I
          DO    20    R = 1,M
            DO    20    LL = 1,L
20               E(LL,R,I) = E(LL,R,I) + C(I,J) * Z(LL,R,J)
```

Step 4: Compute statistics $\hat{cov}(y)$, $\hat{\beta}$ *and* $\tilde{\beta}$

Once we have the three-dimensional array E, we can easily compute estimates such as $\hat{cov}(y)$ defined in (1.4). This equation can also be computed as

$$\hat{cov}(y) = e\ e'/(m-1) - \bar{e}\ \bar{e}'m/(m-1), \tag{2.4}$$

where $\bar{e} = (\bar{e}_1, \ldots, \bar{e}_i, \ldots, \bar{e}_n)'$ with $\bar{e}_i = \Sigma_{r=1}^m e_{ir}/m$. The vectorizable FORTRAN program for the computation of $\bar{e}$ resembles Figure 2. We can also use special FORTRAN 200 instructions such as Q8SSUM, which computes sums like Σe_{ir}. The computation of $\hat{cov}(y)$ in (2.4) can be programmed analogously. Alternatively we can program inner-products ($e'e$ and $\bar{e} \bar{e}'$) through the special function Q8SDOT; see SARA (1984, pp. 22,30).

A problem arises when computing the *inverse* $[\hat{cov}(y)]^{-1}$, which is needed to compute the EGLS estimator $\tilde{\beta}$ in (1.6). The trick in the preceding steps was to make the inner loop long. The instruction within that loop can be executed in parallel, provided that instruction contain

no function or subroutine references except for basic functions such as sine: the vector computer can execute in parallel basic operations only; see SARA (1984, p. 23). So the computer cannot calculate L inverses in parallel, since inversion requires a subroutine call. Hence $[\hat{cov}(y)]^{-1}$ must be computed in scalar mode. Once this inverse is available, some matrix multiplications follow such as $[\hat{cov}(y)]^{-1} X$, which can be vectorized.

3. Computational Tests

Into the OLS estimator of (1.6) we substitute

$$W = (X' X)^{-1} X'$$
(3.1)

and into the EGLS of (1.7) we substitute

$$V = (X' [\hat{cov}(y)]^{-1} X)^{-1} X' [\hat{cov}(y)]^{-1}.$$
(3.2)

W needs to be computed only once, but V is calculated L = 100 times since $\hat{cov}(y)$ changes every time. For these computations we select three cases, as follows. We use a regression model for k factors accounting for all $k(k-1)/2$ two-factor interactions besides the overall mean and the k main effects; so $Q = 1 + k + k(k-1)/2$. The experimental design is a 2^{k-p} design with $n = 2^{k-p} \geq Q$. Hence if $k = 2$ then $Q = 4$ and $n = 2^2 = 4$. If $k = 4$ then $Q = 11$ and $n = 2^4 = 16$. If $k = 6$ then $Q = 22$ and $n = 2^{6-1} = 32$. We keep the number of simulation replicates at its minimum: $m = n + 1$. To improve the accuracy of our timing data we repeat the computation 100 times.

The CYBER 205 can run in vector mode and in scalar mode respectively. For OLS the scalar mode of this expensive computer runs only slightly faster than the VAX does. In vector mode, however, the CYBER takes less than 2% of the VAX time. In our EGLS code, matrix inversion cannot be vectorized, as we saw. Therefore we measure how much time inversion takes. Obviously the scalar mode and the vector mode of the CYBER yield the same CPU times for inversion, apart from measurement errors. In "vector" mode we vectorize all instructions that can be vectorized over the L dimension. In the small problem (n = 4, Q = 4) non-vectorizable inversion takes 85% of total time; consequently, vectorizing the rest can never save more than 15%; it does save 14%. In the large problem (n = 32, Q = 22) inversion takes only 28% of total time; vectorizing the rest saves 70%. Note that for the small problem, EGLS runs faster on the VAX than on the CYBER, even in vector mode.

Part II. Validation and Confidence Interval Procedures: Rao (1959) versus Kleijnen (1983)

To test if the specified regression model is a valid metamodel, we can apply two statistical techniques, developed by Rao (1959) and Kleijnen (1983) respectively. Details are given in Kleijnen (1990).

1. Statistical procedures

Translating Rao's symbols into our notation and assuming that the rank of X is Q leads to the F statistic (which is closely related to Hotelling's statistic):

$$F_{n-Q,m-n+Q} = \frac{m-n+Q}{(n-Q)(m-1)} (\bar{y}-X\tilde{\beta})' \widehat{cov}(\bar{y})^{-1} (\bar{y}-X\tilde{\beta}) = c \, \hat{e}' \widehat{cov}(\bar{y})^{-1} \hat{e}, \qquad (1.1)$$

with constant $c = (m-n+Q)/\{(n-Q)(m-1)\}$ and estimated residuals $\hat{e} = (\bar{y}-X\tilde{\beta})$. Kleijnen (1983) proposes *cross-validation*. For OLS we then get

$$\widehat{var}(\hat{y}_i - \bar{y}_i) = \widehat{var}(\hat{y}_i) + \widehat{var}(\bar{y}_i) - 2 \, cov \, (\hat{y}_i, \bar{y}_i)$$

$$= x_i' \, \widehat{cov}(\hat{\beta}_{(-i)}) \, x_i + \hat{\sigma}_i^2/m - \frac{2}{m} x_i' \, W_{-i} \, \widehat{cov}(y_{-i}, y_i), \qquad (1.2)$$

where $\hat{\sigma}_i^2$ was given in Part 1 and $\widehat{cov}(\hat{\beta}_{(-i)})$ follows from adding the index (-i) to all quantities in (1.3):

$$cov(\hat{\beta}) = (X'X)^{-1} X' (\widehat{cov}(\bar{y})) X (X'X)^{-1}. \qquad (1.3)$$

The $Q\times(n-1)$ matrix W_{-i} in (1.2) is defined by

$$W_{-i} = (X'_{-i}X_{-i})^{-1}X'_{-i},$$ (1.4)

and the vector $\hat{cov}(y_{-i},y_i)$ in (1.2) is defined by

$$\hat{cov}(y_{-i},y_i) = (\hat{\sigma}_{1i},\hat{\sigma}_{2i},\ldots,\hat{\sigma}_{i-1,i},\hat{\sigma}_{i+1,i},\ldots,\hat{\sigma}_{ni})'.$$ (1.5)

For EGLS we replace $\hat{cov}(\hat{\beta}_{(-i)})$ in (1.2) by the analogue of (1.3), again adding the index $(-i)$:

$$cov(\tilde{\beta}) \approx (X'\hat{cov}(\bar{y})^{-1}X)^{-1},$$ (1.6)

and we replace W_{-i} in (1.2) by

$$V_{-i} = (X'_{-i}\,\hat{cov}(y_{(-i)})^{-1}\,X_{-i})^{-1}X'_{-i}\,\hat{cov}(y_{(i)})^{-1},$$ (1.7)

which ignores the random character of $\hat{cov}(y_{(i)})$. These equations are used to compute the standardized prediction error:

$$t_{(i)} = \frac{\hat{y}_i - \bar{y}_i}{\{\hat{var}(\hat{y}_i - \bar{y}_i)\}^{\frac{1}{2}}} \qquad (i = 1,\ldots,n).$$ (1.8)

We assume that (1.8) equals Student's t_v with $v = m-1$, and test this assumption in a Monte Carlo experiment. We use *Bonferroni's* inequality; that is, we test the maximum of the n individual errors $t_{(i)}$ at a significance level α/n (whereas Rao's F statistic is tested at α); we reject the regression model if

$$\max_{1\le i\le n} |t_{(i)}| > t_v^{\alpha/(2n)},$$ (1.9)

where the factor 2 is needed because it is a two-sided test.

From Rao (1959, p. 53) we derive the following $1-\alpha$ two-sided *confidence interval* for the individual regression parameter β_j:

$$\tilde{\beta}_j \pm t_v^{\alpha/2}\,\hat{\sigma}(\tilde{\beta}_j)\left[\frac{1+F\,m(n-Q)/(m-n+Q)}{1-(n-Q)/(m-1)}\right]^{\frac{1}{2}},$$ (1.10)

where $v = (m-1)-(n-Q)$, $\hat{\sigma}(\tilde{\beta}_j) = \{\hat{var}(\tilde{\beta}_j)\}^{\frac{1}{2}}$ with $\hat{var}(\tilde{\beta}_j)$ computed from (1.6), and $F = F_{n-Q,m-n+Q}$ as given by (1.1). To derive (1.10) we must use asymptotic relationships and we interpret Rao; so it seems wise to test its performance in a Monte Carlo experiment.

Eq. (1.1) shows that F↓0 if the regression model fits adequately; if the model does not fit, it makes no sense to derive confidence intervals for the individual parameters; see Rao (1959, pp. 56-57). Further, we suggest to use EGLS only if m is "large" so that $\hat{\sigma}(\tilde{\beta}_j)$ may indeed be computed from the asymptotic relationship in (1.6); for large m we may replace t_v by z. Kleijnen (1988, p. 68) proposes:

$$\tilde{\beta} \pm z^{\alpha/2} \hat{\sigma}(\tilde{\beta}_j). \tag{1.11}$$

Obviously this interval is tighter than Rao's interval (1.10). For the OLS estimator Arnold (1981, p. 343) gives the *exact* interval

$$\hat{\beta}_j \pm t_{m-1}^{\alpha/2} \hat{\sigma}(\hat{\beta}), \tag{1.12}$$

where $\hat{\sigma}(\hat{\beta})$ follows from (1.3). We use a Monte Carlo experiment to examine the confidence intervals (1.10) and (1.11) for EGLS, and (1.12) for OLS.

2. Statistical Design of a Monte Carlo Experiment

We estimate the following *performance measures*.

(i) The type I and type II errors of Rao's F test (based on EGLS), Kleijnen's cross-validation test for OLS, and Kleijnen's test for EGLS.

(ii) The coverage probabilities of the different confidence intervals per individual regression parameter β_j, and the mean interval halfwidths.

The values of the performance measures vary with the *case*, defined by the number of replications m, the covariance matrix cov(y), the design and the true model which determine X, and the regression parameters β. We select the following experimental factors and their levels.

Factor 1, number of simulation replications m: We chose m = n+1, n+10, n+25, and n+50, hoping that as m increases, asymptotic formulas hold.

Factor 2, variance heterogeneity: We quantify the variance heterogeneity through d = max(σ_i)/min(σ_i). We consider only two levels: d=1 (constant variances, so OLS is optimal) and d=10. The magnitude of the variances should be fixed relative to the magnitude of the regression parameters β (see factor 5). So without loss of generality we fix the average standard deviation at the value one. We sample the n-2 intermediate variances uniformly between min(σ_i) and max(σ_i). This yields a unique solution for min(σ_i), namely 2/(1+d). We randomly assign the n standard deviations to the responses that correspond with the n combinations in X.

Factor 3, correlation magnitude: Intermediate results showed that constant and vary-ing correlation coefficients gave the same results (even if the correlation coeffi-cients are constant, their estimates vary). Therefore we report only on the simplest pattern: constant ρ. The magnitude is fixed at three levels: $\rho=0$; 0.5; 0.9.

Factor 4, matrix of independent variables: Most simulation users apply a regression model that falls into one of the following three classes (also see Kleijnen, 1988, p. 69): (a) First order polynomial with main effects β_j (j = 1,...,k) and overall mean β_0. (b) Type (a) augmented with two-factor interactions β_{jg} (j < g and g = 2,...,k). (c) Second order polynomial, which includes quadratric effects β_{jj}.

Let Q_0 and Q_1 denote the number of regression parameters in the user's model and the true model respectively. Then the user applies a validation test with $Q = Q_0$; in the Monte Carlo experiment we generate observations y_{ir} through the true model with $Q = Q_1$. When we estimate the type I error, we make both models coincide; we take the simplest model, that is, we reduce Q_1 to Q_0 (rather than increase Q_0 to Q_1). When estimating the type II error, we make the user's model a subset of the true model ($Q_0 < Q_1$), which is traditional in the experimental design literature.

We consider four levels for this factor. For k=1 the user's model is a first order polynomial. The true model is $E(y_i) = \beta_0 + \beta_1 x_i + \beta_{11} x_i^2$; so $Q_1=3$. The user estimates the two parameters from n = 3 observations. Then the OLS and EGLS estima-tors $\hat{\beta}_{-i}$ and $\tilde{\beta}_{-i}$ coincide, since X_{-i} is a square matrix. For k = 2 we have $E(y_i|H_0)$ = $\beta_0 + \beta_1 x_{i1} + \beta_2 x_{i2}$. So a 2^k design implies n=4. For the estimation of the type II error it is unimportant how many independent variables are ignored; their total effect matters (see factor 5). Therefore we add the two-factor interaction ($\beta_{12} x_1 x_2$) to the first-order model; we do not need to consider a second order model. We also wish to study cases where OLS and EGLS differ. So we augment the 2^2 design (n=4) with the "central" design point ($x_1 = x_2 = 0$): n=5. Moreover we extend the design to a "central composite" design: n=9.

Factor 5, true regression parameters: When we estimate the type I errors of the validation tests, the user's model and to the true model are identical and the mag-nitudes of the regression parameters do not matter. Therefore we take $\beta_j = 0$ (j = 1,...,Q). When we estimate the power of the validation tests, the magnitudes of the ignored regression parameters are important. We select a single ignored parame-ter such that the estimated power exceeds zero but is smaller than one so that the power differences among the various tests become clear. For k=1 we take a quadratic effect of 0.5; for k=2 we select an interaction of 0.5 (remember that $\bar{\sigma} = 1$; as m increases, the power increases). So "factor" 5 is kept constant; it is not really a factor.

Factor 6, the nominal α values: We fix the α value in the validation tests at 0.20: Bonferroni's inequality is conservative so relatively high values are used for α. We fix the α value in the confidence intervals at 0.10. So "factor" 6 is constant.

Altogether the first four factors specify 96 cases (96 = 4 x 2 x 3 x 4). We *replicate* each case L = 100 times in the Monte Carlo experiment (see Part 1). The validation tests give estimated type I and II errors with standard errors that do not exceed 0.05. We compute confidence intervals for β_j, only if the regression model is accepted by the validation test. Consequently coverages and halfwidths are estimated from fewer than 100 replications.

The Monte Carlo experiment is repeated for 8 of the 96 cases, replacing normal errors by lognormal and uniform errors.

The *pseudorandom number generator* is the standard NAG subroutine, which is a multiplicative generator with multiplier 13^{13} and modulus 2^{59}.

3. Monte Carlo Results

Extensive tables are given in Kleijnen (1990). Under normality, different OLS and EGLS estimates in cross-validation do not affect the type I and II errors significantly. Actually we conjectured that EGLS would give better power, if the OLS assumptions do not hold and if the covariance matrix is estimated from many replications m; the intercept estimator $\hat{\beta}_0$, however, is less accurate in EGLS.

Rao's validation test has estimated type I errors not significantly different from the nominal 0.20 value. So our interpretation of Rao and the asymptotic formulas are correct indeed. Cross-validation uses Bonferroni's inequality, which is conservative: the estimated type I errors are smaller than 0.20. A conservative test implies low power: Rao's validation test has higher power.

Positive ρ values do not affect the estimated type I error in cross-validation; so the last term in (1.2) is adequate. Positive correlation does improve the power of the validation tests, so it makes sense to use common seeds in simulation! A high response variance creates so much noise that the power is low, even if ρ is high.

Our EGLS confidence intervals in (1.11) hold only asymptotically: for small m the coverage probability turns out to be too low. The OLS intervals of (1.12) are indeed exact, even for small m. Rao's EGLS confidence intervals of (1.10) have correct coverages; in some cases they are wider than our OLS intervals. For simplicity's sake, however, we may base the confidence intervals on the EGLS point estimates $\tilde{\beta}_j$ that are used in the model validation test.

As ρ increases, the mean halfwidth length decreases , except for the intercept; this phenomenon is explained in Kleijnen (1987, ppp. 172-173).

Under lognormality, EGLS breaks down; only OLS confidence intervals and cross-validation works; see (1.2) and (1.12).

References

ARNOLD, S.F., *The Theory of Linear Models and Multivariate Analysis*, John Wiley & Sons, New York, 1981.

KLEIJNEN, J.P.C., "Cross-Validation using the t Statistic," *European J. Oper. Res.*, 13, 2(1983), 133-141.

KLEIJNEN, J.P.C., *Statistical Tools for Simulation Practitioners*, Marcel Dekker, Inc., New York, 1987.

KLEIJNEN, J.P.C., Analyzing Simulation Experiments with Common Random Numbers, *Management Sci.*, 34, 1(1988), 65-74.

KLEIJNEN, J.P.C., "Pseudorandom Number Generation on Supercomputers," *Supercomputer*, 6 (1989), 34-40.

KLEIJNEN, J.P.C., *Regression Metamodels for Simulation with Common Random Numbers: Comparison of Techniques*, Katholieke Universiteit Brabant (Tilburg University), December 1990. (To be published in *Management Sci.*)

KLEIJNEN, J.P.C. and B. Annink, *Pseudorandom Number Generators for Supercomputers and Classical Computers: a Practical Introduction*, Katholieke Universiteit Brabant (Tilburg University), Oct. 1989. (To be published in *European J. Oper. Res.*)

KLEIJNEN, J.P.C. and B. Annink, *Vector Computers, Monte Carlo Simulation, and Regression Analysis: An Introduction*, Katholieke Universiteit Brabant (Tilburg University), April 1990. (To be published in *Management Sci.*)

LEVINE, R.D., "Supercomputers," *Scientific American*, (1982), 112-125.

RAO, C.R., Some problems involving linear hypotheses in multivariate analysis, *Biometrika*, 46, (1959), 49-58.

SARA, *Cyber 205 User's guide; part 3, Optimization of FORTRAN programs*. SARA (Stichting Academisch Rekencentrum Amsterdam/ Foundation Academic Computer Centre Amsterdam), Amsterdam, 1984.

Appendix: List of Reviewers for this Volume

R. J. Beran
H.H. Bock
U. Dieter
E.J. Dudewicz
D. Edwards
L. Fahrmeir
M. Falk
U. Gather
P. Gänssler
W. Härdle
M. E. Johnson
H.-D. Keller
J.P.C. Kleijnen
W. Krämer
J. P. Kreiß
W. Lehmacher
J. Lehn
I. Pigeot
R.D. Reiss
B.D. Ripley
S. Schach
N. Schmitz
M. Schumacher
B. Streitberg
W. Stute
H.-J. Trampisch
G. Trenkler
G. Tutz
J. Wahrendorf

Lecture Notes in Economics and Mathematical Systems

For information about Vols. 1–210
please contact your bookseller or Springer-Verlag